Springer Theses

Recognizing Outstanding Ph.D. Research

Aims and Scope

The series "Springer Theses" brings together a selection of the very best Ph.D. theses from around the world and across the physical sciences. Nominated and endorsed by two recognized specialists, each published volume has been selected for its scientific excellence and the high impact of its contents for the pertinent field of research. For greater accessibility to non-specialists, the published versions include an extended introduction, as well as a foreword by the student's supervisor explaining the special relevance of the work for the field. As a whole, the series will provide a valuable resource both for newcomers to the research fields described, and for other scientists seeking detailed background information on special questions. Finally, it provides an accredited documentation of the valuable contributions made by today's younger generation of scientists.

Theses are accepted into the series by invited nomination only and must fulfill all of the following criteria

- They must be written in good English.
- The topic should fall within the confines of Chemistry, Physics, Earth Sciences, Engineering and related interdisciplinary fields such as Materials, Nanoscience, Chemical Engineering, Complex Systems and Biophysics.
- The work reported in the thesis must represent a significant scientific advance.
- If the thesis includes previously published material, permission to reproduce this must be gained from the respective copyright holder.
- They must have been examined and passed during the 12 months prior to nomination.
- Each thesis should include a foreword by the supervisor outlining the significance of its content.
- The theses should have a clearly defined structure including an introduction accessible to scientists not expert in that particular field.

More information about this series at http://www.springer.com/series/8790

Jonathan Bortfeldt

The Floating Strip Micromegas Detector

Versatile Particle Detectors for High-Rate Applications

Doctoral Thesis accepted by
the Ludwig-Maximilians-Universität München, Germany

 Springer

Author
Dr. Jonathan Bortfeldt
Fakultät für Physik
Ludwig-Maximilians-Universität München
Garching
Germany

Supervisor
Prof. Otmar Biebel
Fakultät für Physik
Ludwig-Maximilians-Universität München
Munich
Germany

ISSN 2190-5053
Springer Theses
ISBN 978-3-319-18892-8
DOI 10.1007/978-3-319-18893-5

ISSN 2190-5061 (electronic)

ISBN 978-3-319-18893-5 (eBook)

Library of Congress Control Number: 2015939146

Springer Cham Heidelberg New York Dordrecht London

Printed on acid-free paper

Springer International Publishing AG Switzerland is part of Springer Science+Business Media
(www.springer.com)

Supervisor's Foreword

The field of application of detectors for particle radiation extends to the same degree as the detectors can withstand a harsh radiation environment. Though nowadays particle detectors not only detect particles but provide precise information on the location and direction for many particles passing the detector at the same time. Thus the requirements on a detector are high: It should be stable and work reliably. It must remain sensitive for particles under all circumstances. It shall measure all particles' location and direction with the highest precision. Nevertheless, it must be simple and cheap to construct, and easy to operate.

In 1988, microstructured gas detectors had been invented to enhance the spatial resolution beyond that of the classical wire chambers which is limited by diffusion processes and space charge effects. Micromegas, short for Micro-Mesh Gaseous detectors are one kind of microstructured gas detectors consisting of a drift region and a thin gap built coplanar from a finely woven metal mesh and readout electrodes with high voltage applied across for signal amplification. The option to microstructure the readout electrodes into many separate strips allows for enhancing the spatial resolution. The narrow gap of the amplification region guarantees fast charge collection and thus a high rate capability of the detector in principle. Nevertheless, the small size of the gap and the high voltage across it render the Micromegas sensitive to discharges induced by highly ionizing particles. These discharges do not affect the device structure but the recharging time of the high voltage creates a period of insensitivity of the detector. To ameliorate this effect, among several approaches the floating strip type Micromegas detector had been devised where the electric potential of the striplike readout electrodes is allowed to float freely.

In this context the progress in detector development achieved by Dr. Jonathan Bortfeldt is of utmost relevance. By improving the floating strip Micromegas, Dr. Bortfeldt significantly advanced this detector type and thus enabled new fields of applications. His floating strip detector is extremely insensitive to discharge processes caused by highly ionizing particles. Thus the new detectors are exceptionally high rate capable. They can spatially resolve single particles down to a few

10 μm in an enormous flux of up to 7 million particles per square centimeter and second. With a single 6 mm thin plane of his detector Dr. Bortfeldt can also measure the direction of an incident particle with an angular resolution of a few degrees by using the so-called μTPC operation mode. Dr. Bortfeldt developed dedicated floating strip detectors from very thin sheets to cope with both, the high intensity ion beam at the Heidelberg Ion Therapy center HIT and the need of spatial resolution for single particles in the beam. He minimized the amount of material, thus alleviating the impact of multiple scattering on the spatial resolution. The excellent performance of these lightweight detectors will allow for live medical imaging of a patient during ion beam treatment.

Dr. Jonathan Bortfeldt's thesis is a compendium on Micromegas detectors in general and on floating strip detectors in particular. His floating strip Micromegas detectors meet the requirements of easy and cheap to build, robust and simple to operate, excellent spatial resolution and high rate capability. Dr. Bortfeldts detectors, therefore, have a large potential for scientific and medical imaging applications.

Munich, Germany
January 2015

Prof. Otmar Biebel

Abstract

Micromegas are high-rate capable, high-resolution micro-pattern gaseous detectors. Square meter sized resistive strip Micromegas are foreseen as replacement of the currently used precision tracking detectors in the Small Wheel, which is part of the forward region of the ATLAS muon spectrometer. The replacement is necessary to ensure tracking and triggering performance of the muon spectrometer after the luminosity increase of the Large Hadron Collider beyond its design value of 10^{34} cm^{-2}s^{-1} around 2020.

In this thesis a novel discharge tolerant floating strip Micromegas detector is presented and described. By individually powering copper anode strips, the effects of a discharge are confined to a small region of the detector. This reduces the impact of discharges on the efficiency by three orders of magnitude, compared to a standard Micromegas. The physics of the detector is studied and discussed in detail.

Several detectors are developed: A 6.4×6.4 cm^2 floating strip Micromegas with exchangeable SMD capacitors and resistors allows for an optimization of the floating strip principle. The discharge behavior is investigated on this device in-depth. The microscopic structure of discharges is quantitatively explained by a detailed detector simulation.

A 48×50 cm^2 floating strip Micromegas is studied in high energy pion beams. Its homogeneity with respect to pulse height, efficiency, and spatial resolution is investigated.

The good performance in high-rate background environments is demonstrated in cosmic muon tracking measurements with a 6.4×6.4 cm^2 floating strip Micromegas under lateral irradiation with 550 kHz 20 MeV proton beams.

A floating strip Micromegas doublet with low material budget is developed for ion tracking without limitations from multiple scattering in imaging applications during medical ion therapy. Highly efficient tracking of 20 MeV protons at particle rates of 550 kHz is possible. The reconstruction of the track inclination in a single detector plane is studied and optimized. A quantitative description of the systematic deviations of the method is developed, which allows for correcting the reconstructed track inclinations.

The low material budget detector is tested in therapeutic proton and carbon ion beams at particle rates between 2 MHz and 2 GHz. No reduction of the detector uptime due to discharges is observed. The measurable pulse height decreases by only 20 % for an increase in particle rate from 2 to 80 MHz. Efficient single particle tracking is possible at flux densities up to 7 MHz/cm^2. The good multi-hit resolution of floating strip Micromegas is shown.

Acknowledgments

I am indebted to many people who have supported me during the last 3 years. I would like to mention a few, I am truly sorry, if I missed anyone.

First of all, I would like to thank my Doktorvater Prof. Dr. Otmar Biebel for his excellent advice, his continuous support in many fields, the freedom to pursue different ways, and his commitment during the past years.

I want to thank PD Dr. Peter Thirolf for his interest and for writing the second evaluation of this thesis and Profs. Dr. Andreas Burkert and Dr. Gerhard Buchalla for participating in the thesis defense and for making the date possible, despite other commitments.

I am indebted to Dr. Ralf Hertenberger for many excellent ideas and fruitful discussions, not only at four in the morning during test beam or during the coffee breaks, and for his continuous encouragement.

I would also like to thank Prof. Dr. Dorothee Schaile for her uncomplicated support, for the possibility to work on this subject, and for creating a very agreeable atmosphere in the group.

I want to express my gratitude to Prof. Dr. Katia Parodi for the very fruitful and interesting collaboration, opening up many new research fields.

A lot of thanks to Dr. Ilaria Rinaldi for her support during the test measurements at HIT, inspiring discussions, and for proofreading parts of this thesis.

I want to thank my hardware colleagues Alexander Ruschke and André Zibell, for their advice and of course the fun we had together during uncounted breaks, conferences, test beams, and elsewhere. Without their support and help, this thesis would be much thinner.

I thank the whole hardware group, Helge Danger, Bernhard Flierl, Johannes Grossmann, Philipp Lösel, Ralph Müller, Samuel Moll and Elias Pree, especially for their support during the test beams.

I also would like to thank the colleagues from the MAMMA collaboration, for many very inspiring and interesting discussions, especially Dr. Jörg Wotschack,

Prof. Dr. Theodoros Alexopoulos, Dr. Givi Sekhniaidze, George Iakovidis, Stefanos Leontsinis, and Kostas Ntekas.

I want to thank Dr. Ioannis Giomataris for the interesting discussion on floating strip Micromegas detectors.

I thank Dr. Michael Böhmer and Dr. Ludwig Maier for providing the Gassiplex readout and for their generous support and help.

I want to thank Peter Klemm and Attila Varga for their help with the detector construction.

A lot of thanks to all members of the group for the very nice atmosphere, especially Herta Franz and Elke Grimm-Zeidler for the help with bureaucracy, Dr. Stefanie Adomeit, Michael Bender (especially for helping out at tandem), Christopher Bock, Angela Burger, Dr. Philippe Calfayan, Bonnie Chow (especially for her help with my wording struggles), Dr. Günter Duckeck, PD Dr. Johannes Elmsheuser (especially for his always open door), Luis Escobar Sawa, Nikolai Hartmann, Friedrich Hönig, Jasmin Israeli, Dr. Federika Legger, Dr. Jeanette Lorentz, Thomas Maier, Dr. Alexander Mann, Christian Meineck, Christoph Anton Mitterer, Dr. Felix Rauscher, Balthasar Schachtner, Christopher Schmitt, Alberto Vesentini, Dr. Rod Walker, and Josephine Wittkowski.

I would like to thank Dres. Sabine Hemmer, David Heereman, and Marc Otten for their inspiration, for their encouragment, their intelligent advice during the past years, and for their friendship.

Many thanks to my parents Susanne Bortfeldt and Prof. Dr. Dr. Martin Bröking-Bortfeldt, to my siblings Florian Bortfeldt, Dr. Liliane Bortfeldt, and Insa Bortfeldt, and to my parents-in-law Claudia Meinhart and Dr. Walter Meinhart for the CARE packages, Sunday phone-calls, their encouragement, the writing asylum, and many good thoughts and memories.

But above all, I am deeply indebted to my wonderful wife Raphaela Bortfeldt for her incredible support in uncountable ways, her tolerance, her understanding, and her love.

Contents

Chapter 1
Introduction

The development and the performance of novel floating strip Micromegas are discussed in this thesis. Micromegas[1] are high-rate capable, high-resolution micro-pattern gaseous detectors, that are well suited for high-rate particle tracking applications, where good spatial resolution is required. In the following, the underlying ideas and the functional principle of Micromegas are introduced. An introduction to the ATLAS[2] experiment at the Large Hadron Collider at CERN[3] is given and the foreseen upgrade of the outermost shell of the ATLAS detector, the muon spectrometer, with Micromegas is described. Application of Micromegas in medical ion transmission tomography is motivated.

1.1 The Micromegas Detector

In the following, the functional principle of Micromegas is introduced, typical performance and a motivation for ongoing detector research are discussed. References to current applications of Micromegas are given. An introduction to particle detection techniques can be found in Kleinknecht (1992) and in Grupen and Shwartz (2008). Francke and Peskov (2014) give a detailed introduction to the field of micro-pattern gaseous detectors as well as a description of different types, such as Micromegas and GEM detectors or parallel plate avalanche chambers.

---

[1] MICRO-MEsh GAS detector.

[2] A Toroidal LHC ApparatuS.

[3] Conseil Européen pour la Recherche Nucléaire, European Organization for Nuclear Research, Geneva, Switzerland.

© Springer International Publishing Switzerland 2015
J. Bortfeldt, *The Floating Strip Micromegas Detector*, Springer Theses,
DOI 10.1007/978-3-319-18893-5_1

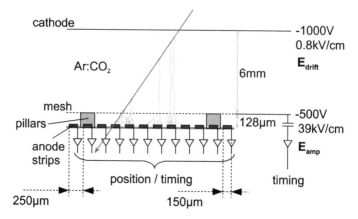

Fig. 1.1 Schematic view of a standard Micromegas with copper anode strips. The micro-mesh is supplied with high-voltage. A charged particle ionizes the detector gas along its path (*blue line*). Ionization electrons move into the high-field region between mesh and anode structure and are amplified in gas avalanches. The resulting charge signals are detected by charge- or current-sensitive preamplifiers

1.1.1 Functional Principle

Micromegas are advanced parallel plate avalanche chambers that were proposed by Giomataris et al. (1996). Gas filled Micromegas consist of a several millimeter wide drift region and an approximately 0.1 mm wide amplification region, separated by a thin conductive micro-mesh. The drift region is formed by a planar metallic cathode and the micro-mesh, the amplification region is defined by the micro-mesh and a highly segmented anode. In the detectors, that have been developed in the course of this thesis, the anode readout structure consists of strips.

A schematic view of a standard Micromegas detector is shown in Fig. 1.1. Traversing charged particles or photons ionize the noble gas based mixture in the drift region of the detector. Depending on the type and the energy of the detected particle, between 50 and 10^5 electron-ion pairs are created. Produced electron-ion pairs are separated by an electric field on the order of 0.8 kV/cm. Ionization electrons drift within 100 ns into the high-field region between mesh and anode strips whereas the ionization ions drift towards the cathode. Upon reaching the amplification region, the electrons are multiplied in an avalanche like process. As the positive ions, created in the gas amplification process, drift towards the micro-mesh, electrons on the anode strips are released and can be detected with charge- or current-sensitive preamplifiers. By reading out every strip individually, the particle hit position as well as the timing can be measured precisely. The interaction processes, the drift of electrons and ions, gas amplification processes and the process of signal formation will be discussed in detail in Chap. 2.

Due to the fast drain of positive ions from gas amplification in about 100 ns, space-charge effects are avoided, that limit the achievable maximum particle rates in gas detectors.

1.1.2 Performance and Challenges

A review of the performance of Micromegas can be found in Beringer et al. (Particle-DataGroup)(2012) and Giomataris (2006). In the following the performance corner stones are summarized.

Due to the finely segmented readout structure with typically four strips per millimeter, Micromegas show a good spatial resolution on the order of 50 μm for perpendicular incident particles. Optimum spatial resolutions below 20 μm have been reached in a detector with ten strips per millimeter using low-diffusion gas mixtures (Derré et al. 2001).

The detection efficiency for traversing charged particles is usually above 95 %. For perpendicularly incident minimum ionizing particles, it is limited by the mesh-supporting structure (Sect. 2.1). The efficiency to typical background radiation like medium energy photons or fast neutrons is low.

Charge amplification factors between a few 10^2 and several 10^4 are typical in Micromegas and can be chosen according to the energy loss and the ionization yield of incident charged particles or photons.

The energy resolution for low energy X-rays is typically on the order of 20 %. Ultimate energy resolutions of 12 % FWHM[4] for 5.9 keV X-rays can be reached in Micromegas with flat, etched copper meshes on a Polyimide holding structure (Galán et al. 2010) or in integrated silicon waver based Micromegas detectors (Chefdeville et al. 2008), which is close to the theoretical limit of 8 % in argon based mixtures (Hashiba et al. 1984).

Strongly ionizing particles can produce large charge densities, that lead to formation of streamers and subsequent discharges between the micro-mesh and the anode structure. The discharges are non-destructive but create dead time due to the necessary restoration of the amplification field. The high-rate capability and the achievable gas amplification factors are limited by the efficiency drop due to discharges. Discharges are intrinsically present in Micromegas. Although their probability can be reduced by the use of light detector gas mixtures with fast ion drift velocities or the addition of preamplification stages, they cannot be completely avoided.

High-rate capable Micromegas development thus aims at reducing the impact of discharges. By adding resistive structures on top of the readout plane, the effect of discharges on the detector performance can be considerably reduced (Alexopoulos et al. 2011; Bilevych et al. 2011).

---

[4]Full Width at Half Maximum.

In this thesis a different approach is chosen. It combines the advantages of standard and resistive strip Micromegas: floating strip Micromegas. Anode strips are individually attached via large quenching resistors to high-voltage, signals are decoupled with small high-voltage resistant capacitors. This concept has been proposed in an early stage by Thers et al. (2001), Bay et al. (2002) and Kane et al. (2003) and is considerably improved in this thesis. The resistance of the recharge resistors is considerably enlarged and coupling capacitances are strongly reduced with respect to previous realizations. This ultimately enables the powerful reduction of the impact of discharges in floating strip Micromegas. The discharge suppression is investigated in detail. Discharges are shown to be limited to a small region of the readout structure. The dead time of the affected region is by two orders of magnitude smaller than in standard Micromegas.

The performance of novel floating strip Micromegas in high-energy pion beams and in low and medium energy, high-rate ion beams is investigated.

1.1.3 Current Applications

A review of the development and the application of micro-pattern gaseous detectors can be found in Dalla Torre (2013). In the following, a few experiments are highlighted, that successfully use Micromegas as tracking detectors or are planning to do so.

The COMPASS[5] experiment at CERN was the first large experiment that used Micromegas as high-rate capable tracking detectors (Abbon et al. (The COMPASS Collaboration) 2007). Before the start of measurements in 2002, extensive R&D studies on Micromegas were performed, optimizing spatial resolution, efficiency and discharge behavior (Thers et al. 2001). COMPASS is a fixed target experiment using high-energy muon and hadron beams. Three stations of four $40 \times 40 \, \text{cm}^2$ Micromegas detectors with strip readout structures provide tracking information at particle hit rates of up to $450 \, \text{kHz/cm}^2$ (Bernet et al. 2005). An upgrade with pixelized Micromegas with resistive layers is proposed for future studies at COMPASS (Thibaud et al. 2014).

T2K[6] is a long baseline neutrino oscillation experiment, sending an intense beam of muon neutrinos over a distance of 295 km through a near detector system, denoted with ND280, to the far detector Super-Kamiokande (Abe et al. 2011). The rectangular Time-Projection-Chambers (TPC), forming the tracking detector of the near detector complex, are read out with Micromegas planes (Abgrall et al. 2011).

The CAST[7] collaboration searches for photon appearance in the high magnetic field of a decommissioned Large Hadron Collider prototype dipole magnet, pointing at the sun, that are produced in axion-to-photon conversion (Zioutas et al. 1999).

---

[5]COmmon Muon and Proton Apparatus for Structure and Spectroscopy.
[6]Tokai-to-Kamioka.
[7]CERN Axion Solar Telescope.

Micromegas are used as low background photon detector for conversion X-rays with a mean energy of 4 keV (Andriamonje et al. 2004). Their good performance in two different configurations is described by Aune et al. (2007) and Galán et al. (2010). Even better background rejection could be achieved with a novel Micro-bulk Micromegas (Dafni et al. 2011).

Micromegas based neutron detector are used in the neutron time-of-flight facility nTOF at CERN for determination of neutron scattering cross-section in different materials. The Micromegas can be equipped with 500 nm thick solid ^{6}Li or ^{10}B converters, charged particles produced in ^{10}B(n,α)^{7}Li or ^{6}Li(n,α)^{3}H reactions and by scattering of neutrons on gas nuclei are detected (Andriamonje et al. 2002; Pancin et al. 2004).

The use of Micromegas detectors is evaluated for the readout of the high-pressure Xenon filled NEXT[8] TPC that will be looking for the neutrinoless double-beta decay in ^{136}Xe (Segui 2013).

The short summary of current or future application demonstrates, that Micromegas are versatile detectors, that can be used in very different applications. Basic R&D on Micromegas detectors, as presented in this thesis, can be very valuable to different fields of physics.

The ATLAS muon spectrometer upgrade with Micromegas is discussed separately in Sect. 1.2.2. Application of Micromegas for imaging in medical ion therapy is treated in Sect. 1.3.

1.2 The LHC, the ATLAS Experiment and Upgrade Plans with Micromegas

The Large Hadron Collider (LHC) at CERN is a high-energy synchrotron designed for proton-proton, lead-lead and lead-proton collisions. The particle beams can be accelerated in two separate beam pipes up to the design center of mass energy of $\sqrt{s} = 14$ TeV with a design luminosity of 10^{34} cm^{-2} s^{-1} for protons and $\sqrt{s} = 1150$ TeV and 10^{27} cm^{-2} s^{-1} for lead ions (Lefevre 2008; Brüning et al. 2004a, b, c).

The LHC commenced operation in 2009. During the proton-proton collisions in Run I, the accelerator was operated at a center of mass energy of 7 TeV in 2010 and 2011 and of 8 TeV in 2012. A maximum luminosity of 7.7×10^{33} cm^{-2} s^{-1} has been reached. Since February 2013 the accelerator system is being upgraded, operation is to be resumed with Run II in early 2015 at a center of mass energy of 13 TeV, the luminosity will gradually be increased to its design value, see Fig. 1.2 for a schedule of data taking runs and upgrades.

Four large experiments are installed at crossing points of the two beams. The ATLAS and the CMS detector[9] (CMS Collaboration 2006) are the two large

---

[8]Neutrino Experiment with a Xenon Time-projection-chamber.
[9]Compact Muon Solenoid.

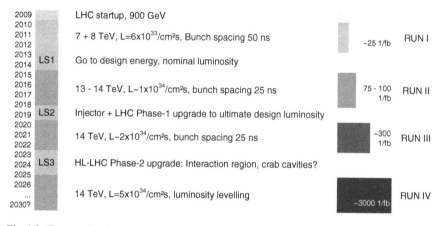

2009	LHC startup, 900 GeV	
2010		
2011	7 + 8 TeV, L=6x10³³/cm²s, Bunch spacing 50 ns	~25 1/fb RUN I
2012		
2013	LS1 Go to design energy, nominal luminosity	
2014		
2015		
2016	13 - 14 TeV, L~1x10³⁴/cm²s, bunch spacing 25 ns	75 - 100 1/fb RUN II
2017		
2018		
2019	LS2 Injector + LHC Phase-1 upgrade to ultimate design luminosity	
2020		
2021	14 TeV, L~2x10³⁴/cm²s, bunch spacing 25 ns	~300 1/fb RUN III
2022		
2023		
2024	LS3 HL-LHC Phase-2 upgrade: Interaction region, crab cavities?	
2025		
2026		
...	14 TeV, L=5x10³⁴/cm²s, luminosity levelling	~3000 1/fb RUN IV
2030?		

Fig. 1.2 Foreseen LHC schedule, beginning with the startup in 2009 up to Run IV with the High-Luminosity LHC. At the moment, the LHC complex is upgraded in the Long Shutdown 1 (LS1) (CERN 2014b). Figure taken from Zibell (2014)

multi-purpose experiments. LHCb[10] investigates CP-violation in the B-meson sector (LHCb Collaboration 1998). ALICE[11] is specifically designed for investigating lead ion collisions in events with very high hit-multiplicity (ALICE Collaboration 1995).

Three smaller detectors are operated in proximity to the larger experiments: TOTEM[12] aims at measuring the total proton-proton cross section (TOTEM Collaboration 2004), LHCf[13] investigates neutral particles in the forward region of ATLAS (LHCf Collaboration 2006) and MoEDAL[14] searches for magnetic monopoles and other exotics (MoEDAL Collaboration 2009).

The ATLAS experiment aims at completing the standard model of particle physics and searches for physics beyond the standard model such as super-symmetry, large extra-dimensions and explanations for the origin of dark matter or the asymmetry between matter and anti-matter in the universe, see (Griffiths 2008) for an introduction to the topic. In 2012 ATLAS and CMS discovered a new particle, that seems to be the long searched for Higgs boson (ATLAS Collaboration 2012; CMS Collaboration 2012). The analysis of its properties is ongoing and will continue as more data is collected, starting in 2015. Up to now all published data is consistent with the expectation for a standard model Higgs boson.

---

[10]LHC beauty.

[11]A Large Ion Collider Experiment.

[12]TOTal Elastic and diffractive cross section Measurement.

[13]LHC forward.

[14]Monopole and Exotics Detector At the LHC.

1.2.1 The ATLAS Experiment

ATLAS is a multi-purpose detector with almost complete angular coverage. A schematic drawing is shown in Fig. 1.3. A comprehensive description of the detector systems can be found in CERN (2014a). Several layers, constructed with different detector technologies, surround the interaction point in the center of the ATLAS detector. The detector consists of three sub-systems, that are shortly described in the following: the inner tracking detector, the calorimeter and the muon system.

The inner detector consists of several layers of high-resolution semiconductor pixel and strip detectors and a straw-tube transition radiation tracker, that allows for electron identification (ATLAS Collaboration 1997a). The complete inner detector is enclosed in a super-conducting 2 T solenoidal magnet. Precisely measuring the curved tracks of charged particles, allows for a determination of their momentum, as the bending radius is a function of particle momentum and charge.

The energy of particles, produced in collisions, is measured in the sampling calorimeter system. It consists of the electro-magnetic lead-liquid argon and the hadronic steel-scintillator calorimeter (ATLAS Collaboration 1996a, b).

The only particles, that are supposed to cross the calorimeter system besides neutrinos, are muons. Muon tracks and momenta are measured in the stand-alone muon spectrometer, that is formed by Monitored Drift Tube and Cathode Strip Chamber tracking detectors and Resistive Plate Chamber and Thin Gap Chamber trigger detectors. Momentum measurement is enabled by bending the muon tracks in toroidal magnetic fields in the barrel and in the two end-cap regions with peak magnetic field

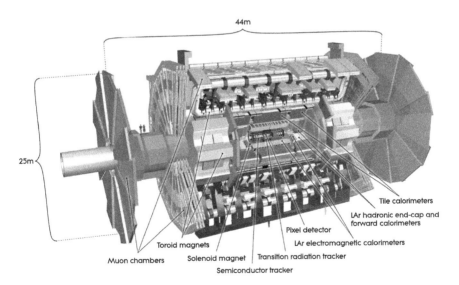

Fig. 1.3 Cut-away view of the ATLAS detector (ATLAS Collaboration 2008)

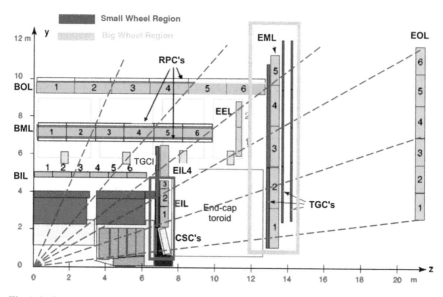

Fig. 1.4 Cut through one quarter of the ATLAS detector. The Monitored Drift Tube muon tracking chambers are marked with *green boxes* in the barrel region and with *cyan boxes* in the end-cap region. A *yellow box* marks the position of the Cathode Strip Chambers. The small wheel and big wheel region are surrounded by *blue* and *yellow rectangles* respectively (ATLAS Collaboration 2013). The chamber naming nomenclature with three letters e.g. BOL is as follows: Barrel chamber are denoted with B, end-cap chambers with E, the three stations are named inner (I), middle (M), outer (O). Two different sizes of chambers exist per station, named small (S) and large (L)

strengths of 3.9 T and 4.1 T respectively. The toroidal fields are produced by super-conducting magnets. The end-cap muon systems consists of three so called wheels: the small wheel directly outside the hadronic end-cap calorimeter and two big wheels at 13 m and 21 m distance from the interaction point, Fig. 1.4 (ATLAS Collaboration 1997b).

1.2.2 New Small Wheel Upgrade

With the upgrade of the LHC luminosity beyond its design value of 10^{34} cm^{-2} s^{-1} in the Long Shutdown 2 in 2018/19 (Fig. 1.2), the background hit rates in the detector systems close to the interaction point and close to the beam pipe increase accordingly. The background consists mainly of low-energy photons and neutrons, background hits in the detectors can be correlated or uncorrelated to triggered collision events.

Elevated background hit rates in the muon detectors in the small wheel region limit the overall ATLAS performance due to two dominant effects (ATLAS Collaboration 2013):

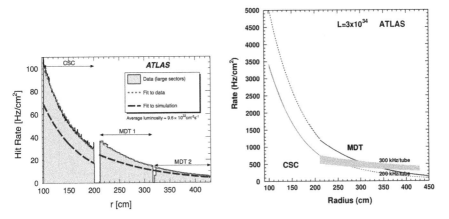

Fig. 1.5 Absolute hit rates in the Cathode Strip Chambers and the Monitored Drift Tube chambers in the small wheel of the ATLAS muon spectrometer (*left*), measured at $\mathcal{L} = 9.8 \times 10^{32}\,\text{cm}^{-2}\,\text{s}^{-1}$ and $\sqrt{s} = 7\,\text{TeV}$. In the *right* figure, the expected hit rates are shown, extrapolated to $\mathcal{L} = 3 \times 10^{34}\,\text{cm}^{-2}\,\text{s}^{-1}$ and $\sqrt{s} = 7\,\text{TeV}$ (ATLAS Collaboration 2013)

1. The performance of the detectors degrades with respect to spatial resolution and efficiency, due to the high occupancy in the muon chambers (Deile et al. 2004). The muon track reconstruction and momentum resolution is significantly degraded.
2. The Level-1 muon trigger in the end-cap region relies at the moment on a muon track angle estimation by Thin Gap Chambers in the middle muon station i.e. the first big wheel, Fig. 1.4. Background hits can be misinterpreted as muon tracks, pointing at the interaction point. In 2012 already 90 % of the Level-1 muon triggers in the end-cap region are due to fake muon tracks.

Currently measured hit rates and extrapolated hit rates after the luminosity upgrade are shown in Fig. 1.5.

The expected rates would significantly distort the behavior of the currently used detectors, such that an upgrade with high-rate capable Micromegas tracking and Thin Gap Chamber triggering detectors is foreseen. Both detector technologies have tracking and triggering capabilities. Details about the foreseen detector technologies and a detailed discussion of the motivation for the upgrade can be found in ATLAS Collaboration (2013).

The research that is presented in this thesis and the developed analysis and reconstruction methods are useful for choosing suitable parameters for the applied Micromegas detectors and are helping in understanding the behavior and the limitations of this detector technology.

Fig. 1.6 Depth dose
distribution for photons and
energy loss distributions for
monoenergetic proton and
carbon ion beams (Fokas
et al. 2009)

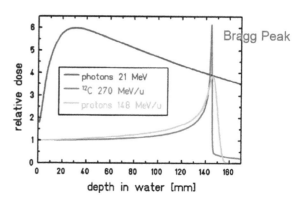

1.3 Medical Application of Micromegas

The use of micro-pattern gaseous detectors such as GEM[15] based detectors (Sauli 1997) for X-ray and ion beam based radiography has been studied by several groups in the past years (Bachmann et al. 2002; Amaldi et al. 2010). Micromegas are well suited for particle tracking applications in medical ion therapy, due to their good spatial resolution, their high-rate capability and their high detection efficiency for charged ions.

In the following, a short introduction to ion beam therapy is given. The Heidelberg Ion Therapy center is introduced. The underlying idea behind ion transmission radiography and tomography is presented.

1.3.1 Ion Beam Therapy and the Heidelberg Ion Therapy Center

A concise introduction to ion beam therapy and its advantages with respect to conventional radiotherapy with photon and electron beams is given by Rinaldi (2011), a detailed review can be found in Fokas et al. (2009). In the following a few aspects are highlighted.

Tumor irradiation aims at damaging tumor cell beyond a point, at which a repair of the cell is possible. Irradiation with ions exploits the specific depth profile of the energy loss and the higher biological effectiveness of heavy charged particles. A comparison of the depth dose distribution of photon, proton and carbon ion beams can be seen in Fig. 1.6.

Due to the rising energy loss of charged particles with decreasing particle energy, charged ions deposit a considerable fraction of their energy in a narrow region, referred to as Bragg peak (Bragg and Kleeman 1905). This enables a more localized

---

[15]Gaseous Electron Multipliers.

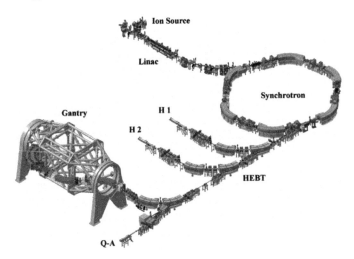

Fig. 1.7 Overview of the accelerator complex (Ondreka and Weinrich 2008). A linear accelerator injects ions with an energy of 7 MeV/u into the synchrotron. Two irradiation places with horizontal beam (*H*1 and *H*2) and a 360° gantry are available for patient treatment. An additional irradiation room (*Q-A*) is used for research and development. A floating strip Micromegas system was tested there

dose deposition in the desired tissue region as compared to photon beams, while sparing healthy or sensitive tissue in the entrance channel and behind the irradiated region.

In order to reach a homogeneous dose deposition in the tumor, the so called three dimensional, intensity controlled raster scanning technique can be used: The particle penetration depth can be varied by adapting the initial particle energy, the beam position can be steered with fast magnets to predefined raster points. Absolute dose deposition under irradiation with a particle beam is controlled over the irradiation time for each raster point.

The accelerator complex at the Heidelberg Ion Therapy center, which has developed out of a pilot project by the GSI Helmholtzzentrum für Schwerionenforschung in Darmstadt (GSI) (Eickhoff et al. 1999) and is Europe's first hospital based facility for tumor irradiation with ions as heavy as carbon, is shown in Fig. 1.7. At the moment proton and carbon ion beams are used for medical irradiation. The available energy range of 48–221 MeV for protons and 88–430 MeV/u for carbon ions corresponds to penetration depths in water between 20 and 300 mm (Ondreka and Weinrich 2008).

1.3.2 Ion Transmission Imaging

Ion beams offer an excellent tumor-dose conformality, due to their inverted depth-dose profile and finite range in tissue. However, they introduce sensitivity to range uncertainties. Imaging techniques play an increasingly important role in ion beam therapy. They support precise diagnostics and identification of the target volume

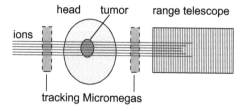

Fig. 1.8 Principle of ion transmission imaging. Ions cross the target volume, their remaining energy is measured in a range telescope. Ion tracking with Micromegas detectors can improve image quality

during treatment planning and ensure the correspondance between planning and actual dose deposition. For improved treatment quality, ion based transmitted images could be acquired at the treatment site before or during treatment and could be employed to monitor the patient positioning. This has the advantage of a by up to two orders of magnitude lower dose deposition as compared to conventional X-ray computed tomography (CT) (Schneider et al. 2004). For a description and comparison of imaging techniques in ion beam therapy, see (Rinaldi 2011).

The principle of ion transmission imaging is shown in Fig. 1.8. Ions with known initial energy traverse the patient, their remaining energy is determined by measuring the residual range of the particles in a suitable telescope (Rinaldi et al. 2013). The energy loss within the target volume is thus determined, which depends on the properties of the transversed tissue, such as e.g. density. The beam position, as given by the raster scanning magnets, is at the moment used for particle localization. In the future, Micromegas tracking detectors in front and behind the patient could allow for single ion tracking and thus a more accurate localization. The image quality is expected to be considerably improved by the additional position information, provided by Micromegas.

High-energy physics detectors were already used in ion range radiography e.g. Pemler et al. (1999); Bucciantonio et al. (2013). Floating strip Micromegas are well suited for high-rate ion tracking due to their good spatial resolution and their low material budget.

The development of an imaging system consisting of floating strip Micromegas and a scintillator based range telescope is ongoing in collaboration with the medical physics group of K. Parodi at LMU. The suitability of floating strip Micromegas for the intended purpose is demonstrated in this thesis.

1.4 On the Content of This Thesis

In this thesis the construction, optimization and performance of floating strip Micromegas is discussed. Reconstruction methods and algorithms that have been developed and implemented for the analysis of the performance in several test campaigns are described in detail. The measurement campaigns are discussed in detail and in a self-contained way, such that readers, interested only in a specific application, can understand the behavior without the need to additionally go through the other test measurements. The overall behavior of floating strip Micromegas in different particle beams and application is discussed and compared.

In the following, the topics, discussed in the different chapters, are mentioned.

- Chapter 2: Functional Principle of Micromegas
  The principle and the underlying processes in Micromegas are discussed. The interaction processes of charged particles and photons in gas detectors, the most probable energy loss, the drift of ions and electron in gases and the gas amplification processes are summarized. Signal formation and the temporal structure of charge signals in Micromegas are described, the typical drift field dependence of the pulse height is explained. The chapter is completed by a qualitative description of discharge formation.

- Chapter 3: Floating Strip Micromegas
  Different types of Micromegas detectors are described, that have been developed during the last years by different groups. The floating strip Micromegas principle is introduced, realization and the improved performance under discharges are discussed. Comparison of results from dedicated discharge studies and a detector simulation allows for explaining the microscopic structure of discharges. Different floating strip Micromegas are described, that were constructed and commissioned in the context of this thesis.

- Chapter 4: Methods
  The applied readout electronics systems and the developed analysis methods and algorithms are described. The signal reconstruction and analysis is discussed, the reconstruction of particle hit positions in the detector is explained. Methods for the determination of spatial resolution and detection efficiency are described, relative detector alignment prescriptions are given. The reconstruction of the inclination of tracks in a single detector plane—the so called μTPC reconstruction—is discussed in detail. Where appropriate, derivations and detailed descriptions have been moved to the appendix.

- Chapter 5: Floating Strip Micromegas Characterization Measurements
  Characterization measurements with standard, with resistive and with floating strip Micromegas are discussed in detail. Calibration measurements with a Micromegas based track telescope in high-energy pion and muon beams are described. The telescope is needed as track reference for studies with a $48 \times 50\,cm^2$ floating strip Micromegas in high-energy pion beams. The observed pulse height behavior, detection efficiency and achievable spatial resolution are discussed, the discharge behavior is described. The high-rate capability and the operation of cosmic muon detecting floating strip Micromegas in high-rate proton background is demonstrated. Irradiation with 20 MeV proton beams is used for an investigation of the ion backflow from the amplification into the drift region in a resistive strip Micromegas.

- Chapter 6: Specific Applications of Floating Strip Micromegas
  Two different applications of floating strip Micromegas are described: Tracking of low energy, high-rate protons in a floating strip Micromegas doublet with

low material budget and ultra high-rate ion tracking and beam characterization measurements at the Heidelberg Ion Therapy center. The performance of the detectors with respect to pulse height, efficiency, spatial resolution and discharge behavior is discussed. The high-rate capability of the detectors is demonstrated, occurring limitations are described.

- Chapter 7: Performance and Properties of Micromegas
  The performance of Micromegas in general and floating strip Micromegas in particular in different particle beams and at different particle rates is discussed. Measured pulse height and efficiency behavior is compared, the observed spatial resolution is described. The single plane angular reconstruction capabilities in different particle beams are discussed, discharge behavior, high-rate capability and its limits are described.

References

Abbon P et al (2007) (The COMPASS Collaboration). The COMPASS experiment at CERN. Nucl Instrum Methods A 577(3):455–518. doi:10.1016/j.nima.2007.03.026. http://www.sciencedirect. com/science/article/B6TJM-4NDMN63-1/2/1dc9387b0a4f19a0c112df2b4f27b539; ISSN 0168-9002

Abe K et al (2011) The T2K experiment. Nucl Instrum Methods A 659(1):106–135. doi:10.1016/ j.nima.2011.06.067. http://www.sciencedirect.com/science/article/pii/S0168900211011910; ISSN 0168-9002

Abgrall N et al (2011) Time projection chambers for the T2K near detectors. Nucl Instrum Methods A 637(1):25–46. doi:10.1016/j.nima.2011.02.036. http://www.sciencedirect.com/science/ article/pii/S0168900211003421; ISSN 0168-9002

Alexopoulos T, Burnens J, de Oliveira R, Glonti G, Pizzirusso O, Polychronakos V, Sekhniaidze G, Tsipolitis G, Wotschack J (2011) A spark-resistant bulk-micromegas chamber for high-rate applications. Nucl Instrum Methods A 640(1):110–118. doi:10.1016/j.nima.2011.03.025. http:// www.sciencedirect.com/science/article/pii/S0168900211005869; ISSN 0168-9002

ALICE Collaboration (1995) ALICE: technical proposal for a large ion collider experiment at the CERN LHC. In: LHC technical proposal CERN-LHCC-95-71, CERN, Geneva. http://cds.cern. ch/record/293391

Amaldi U, Hajdas W, Iliescu S, Malakhov N, Samarati J, Sauli F, Watts D (2010) Advanced quality assurance for CNAO. Nucl Instrum Methods A 617(13):248–249. doi:10.1016/j.nima.2009.06. 087. http://www.sciencedirect.com/science/article/pii/S0168900209013424; ISSN 0168-9002 (Proceedings of the 11th Pisa Meeting on Advanced Detectors)

Andriamonje S, Aune S, Dafni T, Fanourakis G (2004) Ferrer Ribas E, Fischer F, Franz J, Geralis T, Giganon A, Giomataris Y, Heinsius FH, Königsmann K, Papaevangelou T, Zachariadou K (2004) A micromegas detector for the CAST experiment. Nucl Instrum Methods A 518(12):252–255. doi:10.1016/j.nima.2003.10.074. http://www.sciencedirect.com/science/ article/pii/S0168900203028080; ISSN 0168-9002 (Proceedings of the 9th Pisa Meeting on Advanced Detectors)

Andriamonje S, Cano-Ott D, Delbart A, Derré J, íez SD, Giomataris I, González-Romero EM, Jeanneau F, Karamanis D, Leprêtre A, Papadopoulos I, Pavlopoulos P, Villamarín D (2002) Experimental studies of a micromegas neutron detector. Nucl Instrum Methods A 481(13):120–129. doi:10.1016/S0168-9002(01)01246-3. http://www.sciencedirect.com/ science/article/pii/S0168900201012463; ISSN 0168–9002

ATLAS Collaboration (1996a) ATLAS liquid-argon calorimeter: technical design report. In: Technical design report CERN-LHCC-96-041, CERN, Geneva. http://cds.cern.ch/record/331061

ATLAS Collaboration (1996b) ATLAS tile calorimeter: technical design report. In: Technical design report CERN-LHCC-96-042, CERN, Geneva. http://cds.cern.ch/record/331062

ATLAS Collaboration (1997a) ATLAS inner detector: technical design report, 1. In: Technical design report CERN-LHCC-97-016, CERN, Geneva. http://cds.cern.ch/record/331063

ATLAS Collaboration (1997b) ATLAS muon spectrometer: technical Design Report. In: Technical design report CERN-LHCC-97-022, CERN, Geneva. http://cds.cern.ch/record/331068

ATLAS Collaboration (2013) New small wheel technical design report. In: Technical design report CERN-LHCC-2013-006, CERN, Geneva. http://cds.cern.ch/record/1552862

Aune S, Dafni T, Fanourakis G, Ferrer Ribas E, Geralis T, Giganon A, Giomataris Y, Irastorza IG, Kousouris K, Zachariadou K (2007) Performance of the micromegas detector in the CAST experiment. Nucl Instrum Methods A 573(1–2):38–40. doi:10.1016/j.nima.2006.10.249. http://www.sciencedirect.com/science/article/B6TJM-4MCW7M1-N/2/a079fa4a9817845de747535efab42f2a: ISSN 0168-9002 (Proceedings of the 7th International Conference on Position-Sensitive Detectors-PSD-7)

Bachmann S, Kappler S, Ketzer B, Müller Th, Ropelewski L, Sauli F, Schulte E (2002) High rate X-ray imaging using multi-GEM detectors with a novel readout design. Nucl Instrum Methods A 478(12):104–108. doi:10.1016/S0168-9002(01)01719-3. http://www.sciencedirect.com/science/article/pii/S0168900201017193; ISSN 0168-9002 (Proceedings of the ninth Int. Conf. on Instrumentation)

Bay A, Perroud JP, Ronga F, Derré J, Giomataris Y, Delbart A, Papadopoulos Y (2002) Study of sparking in micromegas chambers. Nucl Instrum Methods A 488(1–2):162–174. doi:10.1016/S0168-9002(02)00510-7. http://www.sciencedirect.com/science/article/B6TJM-45BRTKJ-V/2/a284420c8f18198d97bd3ecbd76666d4; ISSN 0168-9002

Beringer J et al (2012) (Particle data group). The review of particle physics. Phys Rev D 86:010001

Bernet C, Abbon P, Ball J, Bedfer Y, Delagnes E, Giganon A, Kunne F, Le Goff J-M, Magnon A, Marchand C, Neyret D, Panebianco S, Pereira H, Platchkov S, Procureur S, Rebourgeard P, Tarte G, Thers D (2005) The $40 \times 40 \, \text{cm}^2$ gaseous microstrip detector Micromegas for the high-luminosity COMPASS experiment at CERN. Nucl Instrum Methods A 536(1–2):61–69. doi:10.1016/j.nima.2004.07.170. http://www.sciencedirect.com/science/article/B6TJM-4D3WCPJ-K/2/b0ba572003c1828607a8023b152aa403; ISSN 0168-9002

Bilevych Y, Blanco Carballo VM, Chefdeville M, Colas P, Delagnes E, Fransen M, van der Graaf H, Koppert WJC, Melai J, Salm C, Schmitz J, Timmermans J, Wyrsch N (2011) Spark protection layers for CMOS pixel anode chips in MPGDs. Nucl Instrum Methods A 629(1):66–73. doi:10.1016/j.nima.2010.11.116. http://www.sciencedirect.com/science/article/pii/S016890021002663X; ISSN 0168-9002

Bragg W, Kleeman R (1905) On the alpha particles of radium and their loss of range in passing through various atoms and molecules. Philos Mag 10:318

Brüning O, Collier P, Lebrun P, Myers S, Ostojic R, Poole J, Proudlock P (eds) (2004a) LHC design report, vol I–the LHC main ring. Report CERN-2004-003-V-1. CERN, Geneva

Brüning O, Collier P, Lebrun P, Myers S, Ostojic R, Poole J, Proudlock P (eds) (2004b) LHC design report, vol II–the LHC infrastructure and general services. Report CERN-2004-003-V-2. CERN, Geneva

Brüning O, Collier P, Lebrun P, Myers S, Ostojic R, Poole J, Proudlock P (eds) (2004c) LHC design report, vol III–the LHC injector chain. Report CERN-2004-003-V-3. CERN, Geneva

Bucciantonio M, Amaldi U, Kieffer R, Malakhov N, Sauli F (2012) Watts D (2013) Fast readout of GEM detectors for medical imaging. Nucl Instrum Methods A 718:160–163. doi:10.1016/j.nima.2012.10.046. http://www.sciencedirect.com/science/article/pii/S0168900212011813; ISSN 0168-9002 (Proceedings of the 12th Pisa Meeting on Advanced Detectors)

CERN (2014a) http://www.atlas.ch/detector.html

CERN (2014b) http://hilumilhc.web.cern.ch

Chefdeville M, van der Graaf H, Hartjes F, Timmermans J, Visschers J (2007) Blanco Carballo VM,
    Salm C, Schmitz J, Smits S, Colas P, Giomataris I (2008) Pulse height fluctuations of integrated
    micromegas detectors. Nucl Instrum Methods A 591(1):147–150. doi:10.1016/j.nima.2008.03.
    045. http://www.sciencedirect.com/science/article/pii/S0168900208004221; ISSN 0168-9002
    (Proceedings of the 9th International Workshop on Radiation Imaging Detectors)
CMS Collaboration (2006) CMS Physics: technical design report vol 1: detector performance and
    software. In: Technical design report CERN-LHCC-2006-001, CERN, Geneva
Collaboration ATLAS (2012) Observation of a new particle in the search for the standard model
    higgs boson with the ATLAS detector at the LHC. Phys Lett B 716(1):1–29. doi:10.1016/j.
    physletb.2012.08.020. http://www.sciencedirect.com/science/article/pii/S037026931200857X;
    ISSN 0370-2693
Collaboration CMS (2012) Observation of a new boson at a mass of 125 GeV with the CMS
    experiment at the LHC. Phys Lett B 716(1):30–61. doi:10.1016/j.physletb.2012.08.021. http://
    www.sciencedirect.com/science/article/pii/S0370269312008581; ISSN 0370-2693
Collaboration ATLAS (2008) The ATLAS experiment at the CERN LHC. JINST 3:S08003
Dafni T, Aune S, Fanourakis G, Ferrer-Ribas E, Galán J, Gardikiotis A, Geralis T, Giomataris I,
    Gómez H, Iguaz FJ, Irastorza IG, Luzón G, Morales J, Papaevangelou T, Rodríguez A, Ruz
    J, Tomás A, Vafeiadis T, Yildiz SC (2011) New micromegas for axion searches in CAST. Nucl
    Instrum Methods A 628(1):172–176. doi:10.1016/j.nima.2010.06.310. http://www.sciencedirect.
    com/science/article/pii/S0168900210015007; ISSN 0168-9002 (Proceedings of the 12th Inter-
    national Vienna Conference on Instrumentation)
Dalla Torre S (2013) MPGD developments: historical roadmap and recent progresses in consoli-
    dating MPGDs. JINST 8(10):C10020. http://stacks.iop.org/1748-0221/8/i=10/a=C10020
Deile M, Dubbert J, Horvat S, Kortner O, Kroha H, Manz A, Mohrdieck-Möck S, Rauscher F,
    Richter R, Staude A, Stiller W (2004) Resolution and efficiency of the ATLAS muon drift-
    tube chambers at high background rates. Nucl Instrum Methods A 535(1–2):212–215. doi:10.
    1016/j.nima.2004.07.193. http://www.sciencedirect.com/science/article/B6TJM-4D4CWC0-4/
    2/98b93af12c149e211c3ef324d5f302ae; ISSN 0168-9002 (Proceedings of the 10th International
    Vienna Conference on Instrumentation)
Derré J, Giomataris Y, Zaccone H, Bay A, Perroud J-P, Ronga F (2001) Spatial resolution
    in micromegas detectors. Nucl Instrum Methods A 459(3):523–531. doi:10.1016/S0168-
    9002(00)01051-2.   http://www.sciencedirect.com/science/article/pii/S0168900200010512;
    ISSN 0168-9002
Eickhoff H, Haberer Th, Kraft G, Krause U, Richter M, Steiner R, Debus J (1999) The GSI cancer
    therapy project. Strahlenther Onkol 175(2):21–24. doi:10.1007/BF03038880. http://dx.doi.org/
    10.1007/BF03038880; ISSN 0179-7158
Fokas E, Kraft G, An H, Engenhart-Cabillic R (2009) Ion beam radiobiology and cancer: time to
    update ourselves. Biochim Biophys Acta-Rev Cancer 1796(2):216–229. doi:10.1016/j.bbcan.
    2009.07.005.   http://www.sciencedirect.com/science/article/pii/S0304419X09000523;   ISSN
    0304-419X
Francke T, Peskov V (2014) Innovative applications and developments of micro-pattern gaseous
    detectors. IGI Global, Hershey
Galán J, Aune S, Carmona J, Dafni T, Fanourakis G, Ferrer Ribas E, Geralis T, Giomataris I, Gómez
    H, Iguaz FJ, Irastorza IG, Kousouris K, Luzón G, Morales J, Mols JP, Papaevangelou T, Rodriguez
    A, Ruz J, Tomás A, Vafeiadis T (2010) Micromegas detectors in the CAST experiment. JINST
    5(01):P01009. http://stacks.iop.org/1748-0221/5/i=01/a=P01009
Giomataris I (2006) High rate applications of micromegas and prospects. ArXiv Physics e-prints,
    Oct 2006
Giomataris Y, Rebourgeard Ph, Robert JP, Charpak G (1996) MICROMEGAS: a high-granularity
    position sensitive gaseous detector for high particle-flux environments. Nucl Instrum Methods A
    376:29–35
Griffiths D (2008) Introduction to elementary particle physics. Wiley, New York
Grupen C, Shwartz B (2008) Particle detectors. Cambridge University Press, Cambridge

Hashiba A, Masuda K, Doke T, Takahashi T, Fujita Y (1984) Fano factor in gaseous argon measured by the proportional scintillation method. Nucl Instrum Methods A 227(2):305–310. doi:10.
1016/0168-9002(84)90138-4. http://www.sciencedirect.com/science/article/B6TJM-473FP27-
21M/2/41bd7fad39b3f5c6f358326d05f1769a; ISSN 0168-9002

Kane S, May J, Miyamoto J, Shipsey I (2003) A study of a MICROMEGAS detector with a new
readout scheme. Nucl Instrum Methods A 505:215–218

Kleinknecht K (1992) Detektoren für Teilchenstrahlung. Teubner

Lefevre C (2008) LHC: the guide (English version). http://cds.cern.ch/record/1092437, Jan 2008

LHCb Collaboration (1998) LHCb: technical proposal. In: LHC Technical proposal CERN-LHCC-
98-004, CERN. http://cds.cern.ch/record/622031

LHCf Collaboration (2006) LHCf experiment: technical design report. In: Technical design report
CERN-LHCC-2006-004, CERN, Geneva. http://cds.cern.ch/record/926196

MoEDAL Collaboration (2009) Technical design report of the MoEDAL experiment. In: Technical
design report CERN-LHCC-2009-006, CERN, Geneva. http://cds.cern.ch/record/1181486

Ondreka D, Weinrich U (2008) The Heidelberg ion therapy (HIT) accelerator coming into operation.
In: Proceedings of the EPAC08, Genoa, Italy, 2008

Pancin J et al (2004) Measurement of the n_TOF beam profile with a micromegas detector. Nucl Instrum Methods A 524(13):102–114. doi:10.1016/j.nima.2004.01.055. http://www.
sciencedirect.com/science/article/pii/S0168900204001512; ISSN 0168-9002

Pemler P, Besserer J, de Boer J, Dellert M, Gahn C, Moosburger M, Schneider U, Pedroni E, Stäuble
H (1999) A detector system for proton radiography on the gantry of the Paul-Scherrer-Institute.
Nucl Instrum Methods A 432(13):483–495. doi:10.1016/S0168-9002(99)00284-3. http://www.
sciencedirect.com/science/article/pii/S0168900299002843; ISSN 0168-9002

Rinaldi I (2011) Investigation of novel imaging methods using therapeutic ion beams. PhD thesis,
Ruprecht-Karls-Universität Heidelberg

Rinaldi I, Brons S, Gordon J, Panse R, Voss B, Jäkel O, Parodi K (2013) Experimental characterization of a prototype detector system for carbon ion radiography. Phys Med Biol 58:413

Sauli F (1997) GEM: a new concept for electron amplification in gas detectors. Nucl Instrum Methods A 386(2–3):531–534. doi:10.1016/S0168-9002(96)01172-2. http://www.sciencedirect.com/
science/article/B6TJM-3SPGW62-3F/2/8fb90f80f092e003152ca74758d2a5c2; ISSN 0168-
9002

Schneider U, Besserer J, Pemler P, Dellert M, Moosburger M, Pedroni E, Kaser-Hotz B (2004) First
proton radiography of an animal patient. Med Phys 31(5):1046–1051. doi:10.1118/1.1690713.
http://scitation.aip.org/content/aapm/journal/medphys/31/5/10.1118/1.1690713

Segui L (2013) NEXT prototypes based on micromegas. Nucl Instrum Methods A 718(0):434–
436. 2012, doi:10.1016/j.nima.2012.10.044. http://www.sciencedirect.com/science/article/pii/
S0168900212011795; ISSN 0168-9002 (Proceedings of the 12th Pisa Meeting on Advanced
Detectors)

Thers D, Abbon Ph, Ball J, Bedfer Y, Bernet C, Carasco C, Delagnes E, Durand D, Faivre
J-C, Fonvieille H, Giganon A, Kunne F, Le Goff J-M, Lehar F, Magnon A, Neyret
D, Pasquetto E, Pereira H, Platchkov S, Poisson E, Rebourgeard Ph (2001) Micromegas
as a large microstrip detector for the COMPASS experiment. Nucl Instrum Methods A
469(2):133–146. doi:10.1016/S0168-9002(01)00769-0. http://www.sciencedirect.com/science/
article/pii/S0168900201007690; ISSN 0168-9002

Thibaud F, Abbon P, Andrieux V, Anfreville M, Bedfer Y, Burtin E, Capozza L, Coquelet C, Curiel
Q, d'Hose N, Desforge D, Dupraz K, Durand R, Ferrero A, Giganon A, Jourde D, Kunne F,
Magnon A, Makke N, Marchand C, D. Neyret, Paul B, Platchkov S, Usseglio M, Vandenbroucke
M (2014) Performance of large pixelised micromegas detectors in the COMPASS environment.
JINST 9(02):C02005. http://stacks.iop.org/1748-0221/9/i=02/a=C02005

TOTEM Collaboration (2004) Total cross-section, elastic scattering and diffraction dissociation at
the large hadron collider at CERN: TOTEM technical design report. In: Technical design report
CERN-LHCC-2004-002, CERN, Geneva. http://cds.cern.ch/record/704349

Zibell A (2014) High-rate irradiation of 15 mm Muon drift tubes and development of an ATLAS compatible readout driver for micromegas detectors. PhD thesis, Ludwig-Maximilians-Universität München

Zioutas K, Aalseth CE, Abriola D, Avignone FT III, Brodzinski RL, Collar JI, Creswick R, Di Gregorio DE, Farach H, Gattone AO, Guérard CK, Hasenbalg F, Hasinoff M, Huck H, Liolios A, Miley HS, Morales A, Morales J, Nikas D, Nussinov S, Ortiz A, Savvidis E, Scopel S, Sievers P, Villar JA, Walckiers L (1999) A decommissioned LHC model magnet as an axion telescope. Nucl Instrum Methods A 425(3):480–487. doi:10.1016/S0168-9002(98)01442-9. http://www.sciencedirect.com/science/article/pii/S0168900298014429; ISSN 0168-9002

Chapter 2
Functional Principle of Micromegas

An introduction to Micromegas has already been given in Sect. 1.1. In the following chapter the functional principle as well as the main features of Micromegas detectors are discussed in detail. Readers interested in the internal setup of Micromegas detectors are referred to Sect. 2.1. The physics of ionization, drift and gas amplification processes are discussed in Sects. 2.2–2.4. In Sect. 2.5 the signal formation in Micromegas is described. Effects related to the transparency of the micro-mesh for electrons are discussed in Sect. 2.6. The streamer mechanism, that leads to the formation of discharges in Micromegas, is shortly described in Sect. 2.7.

2.1 Internal Setup

A Micromegas is a micro-pattern gaseous detector with strongly asymmetric drift and amplification regions. They are defined by three key components: a printed circuit board (PCB), carrying the readout structure, a fine micro-mesh, held by insulating pillars at a small distance of about 0.1 mm to the readout structure and a cathode, which closes the typically 6 mm wide drift region.

The readout structure typically consists of photo-lithographically etched copper strips with a width of 150 μm and a pitch of 250 μm, although more complex patterns like zig-zag-lines, pad or pixels are easily possible (Thibaud et al. 2014). Periodically spaced, insulating solder resist pillars with a diameter of 0.3 mm and a height of 128 μm define the width of the amplification region between anode and micro-mesh. Woven stainless steel meshes[1] have proven to be easy to handle during assembly, to sustain discharges and to permit a satisfactory energy and spatial resolution.

A several millimeter wide drift region is formed by the micro-mesh and a cathode, which can also directly close the gas filled region of the detector. Drift and amplification regions are filled with a suitable detector gas, usually a noble gas based mixture

---

[1] E.g. 400 lines per inch, wire diameter 18 μm, wire periodicity 63.5 μm.

© Springer International Publishing Switzerland 2015
J. Bortfeldt, *The Floating Strip Micromegas Detector*, Springer Theses,
DOI 10.1007/978-3-319-18893-5_2

of e.g. Argon and Carbon Dioxide, Argon and Isobutane (Bay et al. 2002) or Neon, Ethane and Tetrafluoromethane gas (Bernet et al. 2005).

A comparison of different Micromegas types is given in Sects. 3.1 and 3.2. The internal setup of the Micromegas detectors, that have been developed in the context of this thesis, is described in detail in Sect. 3.4.

2.2 Interaction of Particles and Photons in the Detector

The underlying principle of each particle and photon detector is the interaction of the traversing particle or photon with the medium in the active volume of the detector. Since the electro-magnetic cross sections dominate by orders of magnitude over the weak, strong or even gravitational cross sections, five different interaction mechanisms of charged particles are used in gas detectors: Ionization, excitation, production of Čerenkov radiation, bremsstrahlung and production of transition radiation.

The detection mechanism in Micromegas is the ionization of the detector gas by traversing particles or photons. Photons with an energy above several 100 eV deposit energy indirectly over production of charged particles.

Micromegas, as most gas detectors, are usually insensitive to neutral particles such as neutrons or neutral mesons. This improves the detector tracking performance in high-rate neutron background environments such as the ATLAS Small Wheel region (ATLAS Collaboration 2013). If desired, the efficiency to neutrons can be increased by the use of special converter media with high neutron capture cross sections, where the neutron produces a charged particle which is subsequently detected. Thermal neutrons are efficiently detected in ^{3}He-filled detectors via the ^{3}He(n,p)^{3}H+764 keV reaction. In the same way the ^{10}B(n,α)^{7}Li+2.792 MeV reaction is used, where the boron can either be gaseous boron trifluoride or solid elementary boron.

2.2.1 Charged Particles

The interaction of charged particles with matter via ionization, excitation and Čerenkov radiation by exchange of (virtual) photons is described in the Photo Absorption Ionization model (Allison and Cobb 1980). It leads to the Bethe-Bloch formula, which describes the mean energy loss per unit length[2]

$$\left\langle \frac{\mathrm{d}E}{\mathrm{d}x} \right\rangle = -4\pi r_e^2 m_e c^2 \rho N_A \frac{Zz^2}{A\beta^2} \left(\frac{1}{2} \ln \frac{2m_e c^2 \beta^2 \gamma^2 T_{max}}{I^2} - \beta^2 - \frac{\delta}{2} - \frac{C}{Z} \right), \quad (2.1)$$

---

[2] r_e: classical electron radius, m_e: electron mass, ρ: density of the target material, N_A: Avogadro constant, Z: atomic number and A: atomic mass in g/mol of target material, z: charge of incident particle, $\beta = v/c$: velocity of incident particle, $\gamma = 1/\sqrt{1-\beta^2}$: Lorentz factor, $T_{max} \approx 2m_e c^2 \beta^2 \gamma^2$: maximum kinetic energy transferable to an electron in an elastic collision, I: mean excitation energy.

Fig. 2.1 Mean energy loss of a singly charged particle in an Ar:CO$_2$ 93:7 gas mixture at 20°C and 1013 mbar in the range $0.1 \leq \beta\gamma \leq 2000$. The Barkas-Berger shell corrections for small $\beta\gamma$ and radiative corrections i.e. bremsstrahlung for large $\beta\gamma$ have been neglected

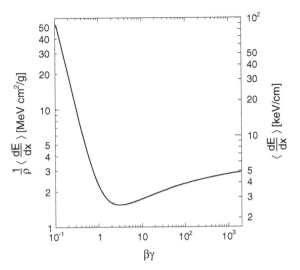

with the density and atomic structure corrections in the Sternheimer-Peierls and Barkas-Berger parametrization respectively. Tabulated values for the density correction δ and the mean excitation energy I for various materials can be found in Sternheimer et al. (1984). For compounds or mixtures the mean energy loss can be calculated from the weight fraction w_j of the jth component

$$\frac{1}{\rho} \left\langle \frac{dE}{dx} \right\rangle = \sum_j \frac{w_j}{\rho_j} \left\langle \frac{dE}{dx} \right\rangle_j . \tag{2.2}$$

The mean energy loss of a singly charged particle in an Ar:CO$_2$ 93:7 vol% gas mixture at normal temperature and pressure, calculated from Eqs. (2.1) and (2.2), is shown in Fig. 2.1. Its universal shape is best visible by plotting it as a function of $\beta\gamma$. The mean energy loss decreases approximately like $1/\beta^2$ for $\beta\gamma \lesssim 1$ and has a minimum with only small variations for $2 \lesssim \beta\gamma \lesssim 8$. Particles with a velocity in this region are called minimum ionizing. For almost all materials, the mean energy loss of minimum ionizing particles is on the order of 2 MeV cm^2/g. For increasing $\beta\gamma$ the mean energy loss rises like $\ln\left(\beta^2\gamma^2\right)$ and reaches the so called Fermi plateau for $\beta\gamma \gtrsim 500$.

It has been argued by Bichsel (2006), that the mean energy loss $\langle dE/dx \rangle$, described by Eq. (2.1), is not directly accessible by measurements in thin detectors, since large energy transfers to target electrons strongly influence the measured values. In particle identification detectors, that use the repeated measurement of energy loss, methods like the so called truncated mean have been applied to estimate the mean energy loss, where the set of highest measured values is discarded. It is discussed by Beringer et al. (Particle Data Group) (2012) to rather use the most probable energy loss $\Delta E/\Delta x|_p$,

Fig. 2.2 Most probable
energy loss of a singly
charged particle in an
Ar:CO$_2$ 93:7 gas mixture at
20 °C and 1013 mbar for gas
volumes with different
thickness d

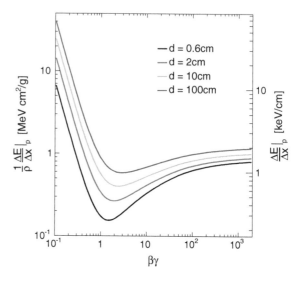

formulated in Landau's theoretical description of ionization in gas detectors. For
charged particles in a thin detector with thickness d we get

$$\left.\frac{\Delta E}{\Delta x}\right|_p = 4\pi r_e^2 m_e c^2 N_A \frac{Zz^2\rho}{A\beta^2}\left(\frac{1}{2}\ln\frac{2m_e c^2\beta^2\gamma^2\xi}{I^2}+\frac{j}{2}-\frac{\beta^2}{2}-\frac{\delta}{2}\right),\quad (2.3)$$

where $\xi = 2\pi r_e^2 m_e c^2 N_A \frac{Zz^2\rho d}{A\beta^2}$ and $j = 0.200$ (Bichsel 1988; Grupen and Shwartz
2008). It should be noted, that the most probable energy loss $\Delta E/\Delta x|_p$ depends,
unlike the mean energy loss $\langle dE/dx\rangle$, on the thickness d or ρd of the sensitive
volume. For increasing thickness, it approaches the mean energy loss, described by
the Bethe-Bloch formula. In Fig. 2.2 the most probable energy loss for an Ar:CO$_2$
93:7 vol% gas mixture, calculated from Eqs. (2.2) and (2.3), is shown.

Although the straggling of the most probable energy loss in thin gas detectors
is considerably larger than predicted by the Landau-Vavilov-Bichsel theory i.e. the
measured pulse height spectra are not adequately described by a Landau function, the
maximum of the distribution is correctly described by Eq. (2.3). In order to extract
the measured most probable pulse height from the distribution of pulse heights,
the distribution is fitted with a Landau function, convoluted with a Gaussian. The
straggling can thus be absorbed into the Gaussian function.

The ionization yield i.e. the total number of produced electron-ion pairs can then
be calculated from the mean or most probable energy loss ΔE in the active area of
the gas detector

$$n_{\text{tot}} = \frac{\Delta E}{W_I},\quad (2.4)$$

Table 2.1 Properties of typical elementary and molecular detector gases at 20 °C and 1013 mbar (Beringer et al. (Particle Data Group) 2012; Sternheimer et al. 1984; Grupen and Shwartz 2008)

Gas	Z	Z/A (g/mol)	ρ (g/cm^3)	I (eV)	W_I (eV)	n_p (cm^{-1})	n_t (cm^{-1})
He	2	0.49967	1.79×10^{-4}	41.8	41.3	3.5	8
Ne	10	0.49556	8.3851×10^{-4}	137.0	37	13	40
Ar	18	0.45059	1.6620×10^{-3}	188.0	26	25	97
Kr	36	0.42959	3.4783×10^{-3}	352.0	24	22	192
Xe	54	0.41130	5.4854×10^{-3}	482.0	22	41	312
CO_2	–	0.49989	1.8421×10^{-3}	85.0	34	35	100
CH_4	–	0.62334	6.6715×10^{-4}	41.7	30	28	54
C_2H_6	–	0.59861	1.2532×10^{-3}	45.4	26	48	112

Z atomic number, only for elementary gases, Z/A (average) atomic number/atomic mass, ρ density, I mean excitation energy, W_I average energy per created electron-ion pair, n_p and n_t mean number of primary and total number of electron-ion pairs created by a minimum ionizing particle of unity charge

where W_I is the average energy, necessary for the production of one electron-ion pair. W_I is for most materials more than twice as high as the ionization potential, since energy is lost due to excitation of the gas atoms or molecules. It is, similar to the energy loss, Bragg additive, such that it can be calculated for mixtures from a relation similar to Eq. (2.2). Higher energetic secondary particles or photons, leaving the detector active volume without depositing their full energy, can complicate the relation between the calculated energy loss and the detectable ionization yield.

A compilation of typical detector gases and their properties can be found in Table 2.1.

Čerenkov radiation is emitted by relativistic charged particles in media with refractive index n, when their velocity exceeds the speed of light in the medium

$$\beta \geq \frac{1}{n} . \tag{2.5}$$

The emission can be understood as the coherent sum of the electric fields of molecular dipoles, created by the polarization of the medium by the traversing particle. As the particle is faster than the speed of light, the sum is non-zero (Čerenkov 1937).

Čerenkov radiation is emitted under a specific angle

$$\cos \vartheta_{\check{c}} = \frac{1}{\beta n} \tag{2.6}$$

to the particle direction of flight. The number of emitted photons can be calculated from the Frank-Tamm formula (Grupen and Shwartz 2008) and is for the visible range between $\lambda = 400$ and 700 nm given by

$$\frac{dN}{dx} = 490 \sin^2 \vartheta_{\check{c}} \ \text{cm}^{-1} . \tag{2.7}$$

The emission of a high-energetic photon, due to the deceleration of a particle with an energy E above the so called critical energy in the proximity of a nucleus, is called bremsstrahlung. Although bremsstrahlung in gases is encountered in e.g. the interaction of electrons and positron with residual gas molecules and ions in storage rings and imposes additional radiation protection measures (Ipe and Fassò 1994), the cross sections in gases are too low to be relevant in gas detectors. The radiation length for Argon at normal temperature and pressure is with $X_0 = 120\,m$ much larger than typical detector dimensions.

Transition radiation with an energy of several keV is emitted by a highly relativistic particle traversing a boundary with alternating refractive indices. This can be understood as the emission of radiation from the variable dipole, formed by the incoming particle with its mirror charge. In practice, several layers consisting of e.g. carbon or Mylar are stacked to increase the photon yield (Beringer et al. (Particle Data Group) 2012). Due to interference effects, the transition radiation yield saturates at certain values. Transition radiation, produced in a radiator between the straw tubes in the Xenon filled ATLAS Transition Radiation Tracker, is used for electron/pion separation (Sect. 1.2.1). There are ideas to use Xenon filled Micromegas detectors, covered with a suitable radiator, in the same way.

2.2.2 Photons

Unlike the interaction of particles with matter, the interaction of photons, collimated to a narrow beam, is mostly a binary process. A photon either interacts with matter and is lost or re-emitted at lower energy or scattered out of the beam or it does not interact.

The intensity of a mono-energetic narrow photon beam with initial intensity I_0 after traversing a material with thickness d is given by

$$I(x) = I_0 \exp(-\mu d) = I_0 \exp(-\frac{\mu}{\rho}x) \,, \qquad (2.8)$$

where ρ is the density, $x := d\rho$ the mass thickness and μ/ρ the mass attenuation coefficient of the material. The energy dependent mass attenuation coefficient is related to the interaction cross section $\sigma(E)$ via

$$\frac{\mu(E)}{\rho} = \frac{\sigma(E)}{m} \,, \qquad (2.9)$$

where m is the atomic mass of the absorber material.

For mixtures and compounds, an effective mass attenuation coefficient can be calculated

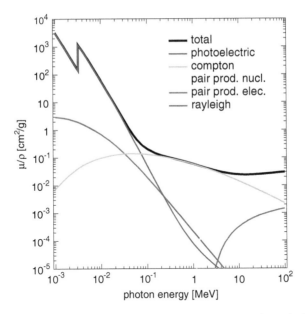

Fig. 2.3 Mass attenuation coefficient as a function of the photon energy for an Ar:CO$_2$ 93:7 vol% gas mixture. The data are taken from Deslattes et al. (2010). Photoelectric effect, Compton scattering and nuclear pair production dominate the mass attenuation coefficient for low, medium and high photon energies. Coherent Rayleigh scattering of photons on electrons, i.e. without excitation or ionization of the material and pair production in the electron field, add small contributions for low and high photon energies respectively

$$\frac{\mu}{\rho}_{\text{eff}} = \sum_i w_i \frac{\mu}{\rho}_i \ , \qquad (2.10)$$

with the weight fraction w_i of the ith component with coefficients μ/ρ_i.

Three main processes dominate the underlying photon interaction with matter: Photoelectric effect, Compton scattering and electron-positron pair production (Deslattes et al. 2010), see Fig. 2.3.

Photoelectric effect, the absorption of a photon with energy E_γ by an atom, accompanied by the release of a shell electron, dominates the interaction cross section for low photon energies up to several 100 keV. The emitted electron energy is given by

$$E_{e^-} = E_\gamma - E_b \ , \qquad (2.11)$$

where E_b is the binding energy of the electron. If the emitted electron comes from a lower shell, the atom is left in an excited state and will typically deexite by emission of low energy photons with total energy $\sim E_b$. If a photon interacts via photo-effect in the active volume of a gas detector, and the active region is thick enough to also detect the photons from the atom deexcitation, the whole incident photon energy is deposited in the detector.

The photoelectric cross section strongly depends on the atomic number Z of the material

$$\sigma_{photo} \propto Z^5 , \tag{2.12}$$

such that high-Z materials like Xenon can be used in gaseous photon detectors to considerably increase the detection efficiency (Kleinknecht 1992).

Compton effect can be interpreted as elastic scattering of photons on quasi free electrons. From energy and momentum conservation, the energy of the scattered photon can be calculated as

$$E'_\gamma = \frac{E_\gamma}{1 + (1 - \cos \vartheta) E_\gamma / m_e c^2} . \tag{2.13}$$

Thus the scattered electrons possess a maximum kinetic energy

$$T'_{e,max} = E_\gamma \frac{2 E_\gamma / m_e c^2}{1 + 2 E_\gamma / m_e c^2} , \tag{2.14}$$

which leads to the so called Compton edge in the energy spectrum.

Photons with $E_\gamma \geq 2 m_e c^2$ can create an electron-positron pair in the proximity of a nucleus. This process is called nuclear pair production. The nucleus is necessary for simultaneous energy and momentum conservation. The analogous process in the proximity of an electron is accordingly called electron pair production.

2.2.3 Multiple Scattering of Charged Particles

Due to the numerous but small interactions of charged particles with the detector components and the detector gas, the particle direction of flight is altered. This effect is called multiple scattering and is typically unwanted in particle detectors. It leads to an increased divergence of a particle beam with initially parallel particle momenta.

The width θ_0 of the Gaussian shaped angular distribution, resulting from multiple scattering of a narrow beam of particles with parallel momenta p and charge number z in a material of thickness x and radiation length X_0 is given by Beringer et al. (Particle Data Group) (2012)

$$\theta_0 = \frac{13.6 \, \text{MeV}}{\beta c p} z \sqrt{\frac{x}{X_0}} \left(1 + 0.038 \ln \left(\frac{x}{X_0} \right) \right) . \tag{2.15}$$

The angular scattering can be translated into a transverse broadening of the particle beam. Its width after a distance d to the scattering object is approximated by

$$\Delta y = \frac{\theta_0}{\sqrt{3}} d. \tag{2.16}$$

For a composite scatterer, consisting of layers of materials with different radiation lengths X_i and densities ρ_i and thickness d_i, the resulting radiation length can be calculated:

$$\frac{1}{X_0} = \sum \frac{w_i}{X_i}, \tag{2.17}$$

where the weight fraction $w_i = d_i \rho_i / \sum d_j \rho_j$ (Gupta 2013).

Multiple scattering distorts the measured spatial resolution in Micromegas detectors, as often straight tracks are assumed as prerequisite. For many applications of Micromegas in low and medium energy particle tracking, the detector material budget is reduced wherever possible, to decrease the degradation of the particle beam.

2.3 Drift of Electrons and Ions in Gases

The drift of electrons and charged ions in gases is governed by electric and magnetic fields $\vec{E}$ and $\vec{B}$. The drift velocity for electrons is given by

$$\vec{v}_d = \frac{e}{m_e} \frac{\tau}{1+\omega^2\tau^2} \left(\vec{E} + \frac{\omega\tau}{B}(\vec{E} \times \vec{B}) + \frac{\omega^2\tau^2}{B^2}(\vec{E} \cdot \vec{B})\vec{B} \right), \tag{2.18}$$

where e and m_e are electron charge and mass, respectively, $\omega = eB/m_e$ is the Larmor frequency and τ the mean time between two collisions with gas atoms (Beringer et al. (Particle Data Group) 2012).

Three properties of the electron drift can be deduced from Eq. (2.18).

First, in the absence of a magnetic field, the electrons follow on average the electric field lines. Electron diffusion transverse and parallel to the electric field, not considered in the equation, leads to a broadening of an initially point-shaped charge distribution.

Second, for perpendicular electric and magnetic fields, the $\vec{E} \times \vec{B}$-term leads to a drift component perpendicular to the fields, such that the net drift velocity vector spans the so called Lorentz angle $\alpha_L \approx \arctan(\omega\tau)$ with the electric field vector. This effect leads to a systematic shift of hit positions with perpendicularly incident particles and compresses or disperses the detected charge distribution for inclined tracks in Micromegas used in a magnetic field.

Third, for non perpendicular electric and magnetic field, there is a drift component into the direction of the magnetic field. For large magnetic fields, i.e. $\omega\tau \gg 1$ the net drift vector may even point into the same direction as $\vec{B}$. Diffusion of electrons perpendicular to the magnetic field is strongly suppressed, as the field forces the electrons on a helix around the magnetic field lines. In Time-Projection-Chambers, this is used to improve the spatial resolution, which would suffer from large electron diffusion.

In the following, the magnetic field is assumed to be zero.

Fig. 2.4 Electron drift velocity as a function of the electric drift field for various Ar:CO$_2$ gas mixtures at 20 °C and 1013 mbar, computed with MAGBOLTZ (Biagi 1999). The color coding is explained in Fig. 2.5

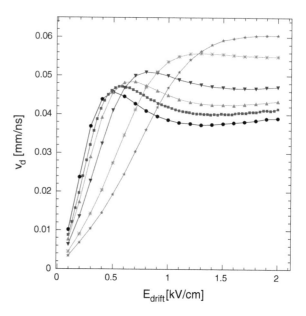

Due to their small mass, electrons gain sufficient energy in the electric field such that their de-Broglie-wavelength is of the same order as typical atomic dimensions, which leads to a strong energy and electric field dependence of their interaction with the gas.

In practice, electron drift velocities as well as diffusion coefficients as a function of the electric field are calculated with MAGBOLTZ (Biagi 1999), see Figs. 2.4 and 2.5. Usually mixtures of gases are used in order to optimize the detector performance. In this thesis gas mixtures are defined by the volumetric mixing ratios of the constituent gases.

The drift of ions can be more easily described by

$$v_{d,\text{ion}} = \mu_{\text{ion}} E \, , \tag{2.19}$$

where μ_{ion} is the gas dependent ion mobility, which is constant over a wide electric field range and not strongly influenced by the admixture of molecular gases. Mobilities for several ions in gases are compiled in Table 2.2.

The drift of ions from gas amplification from the anode strips towards the mesh defines, together with the electron drift time in the drift region, the intrinsic length of signals in Micromegas. In order to further improve the high-rate capabilities by decreasing the signal length, the use of gases lighter than Argon is necessary. This has furthermore an advantageous influence on the discharge probability, as a quick drain of space charge reduces the streamer formation probability, see Sect. 2.7.

Fig. 2.5 Transverse diffusion coefficient as a function of the electric drift field for various Ar:CO$_2$ gas mixtures at 20 °C and 1013 mbar, computed with MAGBOLTZ. The volumetric mixing ratios are stated in the legend. From variation of the Monte Carlo integration parameters, used in MAGBOLTZ, an accuracy for the shown coefficients of the order 5 % is approximated

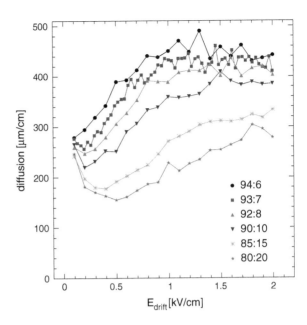

Table 2.2 Mobility of ions in gases at 20 °C and 1013 mbar (MacDaniel and Mason 1973)

Gas	Ion	μ_{ion} [cm^2/(Vs)]
He	He$^+$	10.4
Ne	Ne$^+$	4.7
Ar	Ar$^+$	1.54
Ar:CH$_4$	CH$_4^+$	1.87
Ar:CO$_2$	CO$_2^+$	1.72
CH$_4$	CH$_4^+$	2.26
CO$_2$	CO$_2^+$	1.09

2.4 Gas Amplification

The ionization charge, created by minimum ionizing particles in gas detectors, is on the order of only 100 e/cm, Table 2.1. Thus in order to reliably detect the passage of singly charged particles or photons, a process of charge amplification is needed. In the following, charge amplification in gas detectors by electron impact ionization, secondary emission and Penning transfer is discussed.

Free electrons from e.g. ionization processes, accelerated by an appropriate electric field, can gain a sufficient kinetic energy between two collisions with gas atoms to further ionize the gas atoms or molecules by impact ionization.

Fig. 2.6 First Townsend
coefficient as a function of
the amplification field for
different Ar:CO$_2$ gas
mixtures at 20 °C and
1013 mbar, calculated with
the program MAGBOLTZ
(Biagi 1999). Note that
Penning transfer is not
incorporated in the shown
values

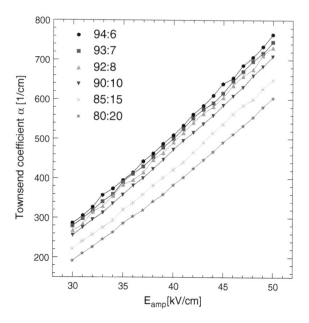

Assuming n_0 electrons entering the high electric field region between mesh and
anode strips of a Micromegas detector, the total number of electrons after a path
length x is given by

$$n(x) = n_0 \exp(\alpha x) . \tag{2.20}$$

The first Townsend coefficient α is a function of the electric field E, the gas
density and the gas type. Calculated Townsend coefficients for several gas mixtures
are shown in Fig. 2.6.

The so called gas gain is the factor between the initial and the final charge and is
defined as

$$G = \exp(\alpha d) , \tag{2.21}$$

where d is the total width of the amplification region.

The density of the detector gas considerably influences the gas gain. For Argon
based mixtures, the relation between gas density n, temperature T and pressure p is
given by the ideal gas law

$$n = \frac{p}{k_B T} , \tag{2.22}$$

where k_B is the Boltzmann constant.

In measurements with a standard Micromegas under controlled gas temperature and pressure conditions by Lippert (2012), supervised by the author, the gas density dependence of the gas gain (Eq. 2.21) has been investigated. The first Townsend coefficient can be parametrized with an empirical relation by Townsend (1910)

$$\alpha = A_0 n \exp\left(-\frac{B_0 n}{E_{\text{amp}}}\right) , \tag{2.23}$$

where $A_0 = k_B A$ and $B_0 = k_B B$ are gas dependent parameters. For an Ar:CO$_2$ 93:7 vol% gas mixture, mean parameters of $A = (111.2 \pm 0.6)$ K/(bar µm) and $B = (2196 \pm 7)$ K V/(bar µm) have been found (Lippert 2012).

If the electric field in the amplification region is sufficiently large, impact ionization by positive ions or photo-effect by photons, produced in gas de-excitation, can lead to secondary emission of electrons from the mesh. This is in Micromegas typically only relevant, if the amplification field is considerably enhanced during discharge formation.

Şahin et al. (2010) state, that the observed gas gains in several gas mixtures are significantly higher than the value predicted by the Townsend theory, due to Penning like transfers i.e. the transformation of excitation energy of one to ionization of the other gas constituent. Penning and Penning-like transfers are possible, if constituent A can be excited to energy levels higher than the ionization threshold of B. Several transfer processes are possible for binary mixtures with constituents A and B, see Kuger (2013).

Deexcitation processes of A^*, competing with Penning transfers, are inelastic excitation of B to energy levels below the ionization energy e.g. excitation of rotational or vibrational energy levels, or radiative decays of A^* and excimers via photons, leaving the active detector area.

The contribution of Penning transfers to the overall gas gain can be quantified by the transfer rate r, describing the fraction of excited atoms or molecule of species A in states, eligible for direct transfer, that lead to an ionization of B. Şahin et al. (2010) claim a transfer rate for an Ar:CO$_2$ 93:7 vol% of (0.42 ± 0.03). Upon comparing gas gain measurements by Kuger (2013) and the author, with a GARFIELD++ simulation (Veenhof 2010) incorporating Penning transfers with the stated transfer rate, a good agreement between measured and calculated gas gains is shown. Measurements conducted by Moll (2013) with strongly ionizing alpha particles in a standard, non-floating-strip Micromegas, confirm the correct description of gas amplification by the GARFIELD++ program, with the Penning transfer rate, stated above.

Typical gas gains between 600 and 10^4 are used in Micromegas.

The measurement of absolute gas gains in Micromegas is possible with two different methods:

1. For low rate irradiation, a measurement of the pulse height of charge signals on the anode with charge sensitive preamplifiers allows for a determination of the gas gain. Due to the considerable fluctuations in energy loss of charged particles in gas detectors and the dependence of the most probable energy loss on the

exact drift gap width (Sect. 2.2.1), energy deposition by low energy X-rays yields more reliable results. Low energy photons e.g. 5.9 keV X-rays, produced in the β-decay of ^{55}Fe, interact via photo-effect with the detector gas (Sect. 2.2.2). The produced photoelectron deposits its kinetic energy within 0.5 mm and produces on average an ionization charge of 5.9 keV/26.56 eV=222 e = q_0. Considering the electron mesh transparency $t(E_{drift})$, the charge-to-voltage conversion factor of the preamplifier $c_{q \to U}$ and a detector capacitance dependent factor c_{cap} (see Bortfeldt 2010, Chap. 8 for a discussion), the pressure and temperature dependent gas gain $G(E_{amp}, T, p)$ can be calculated from the measured pulse height $ph(E_{amp}, E_{drift}, T, p)$ from

$$G(E_{amp}, T, p) = \frac{ph(E_{amp}, E_{drift}, T, p)}{q_0 t(E_{drift}) c_{cap} c_{q \to U}}. \tag{2.24}$$

This method relies on the accurate knowledge of the preamplifier specific factor $c_{q \to U}$ and the detector and readout electronics specific factor c_{cap}.

2. At high-rate irradiation with charged particles or low energy photons, a constant current of ionization charge is produced. Measuring the resulting current between anode and mesh $I(f, E_{drift}, E_{amp}, T, p)$ allows for an alternative gas gain determination. The gas gain

$$G(E_{amp}, T, p) = \frac{I(f, E_{drift}, E_{amp}, T, p)}{q(E_{drift}) f} \tag{2.25}$$

depends on the particle rate f and the charge, reaching the amplification region $q(E_{drift}) = q_0 t(E_{drift})$, which is given by the ionization yield q_0 and the drift field dependent electron mesh transparency $t(E_{drift})$. This method is insensitive to preamplifier related calibration, such as the charge-to-voltage conversion and detector capacitance correction factor as well as preamplifier shaping time. For low energy particles and long particle tracks in the detector, the predicted mean energy loss can be used to calculate the ionization yield. Care must be taken, that high-rate effects such as saturation, anode charge-up, ion space-charge in the amplification region or ion backflow corrections are taken into account.

Absolute gas gain factors are inherently difficult to measure with Micromegas detectors: The pulse height of signals, measured with charge sensitive preamplifiers at the anode, is strongly influenced by the internal capacitances within the detector and less strongly by the charge-to-voltage conversion factors of the preamplifiers (Bortfeldt 2010; Bortfeldt et al. 2012). In the alternative continuous current mode, where the constant current between mesh and anode strips, caused by continuous ionization of a high-rate particle or photon beam is measured, these calibration effects can be avoided. They are unfortunately replaced by distortions of the gas gain due to the charge-up, ion backdrift and other high-rate effects.

In this thesis, the absolute gas gain has been determined e.g. with 20 MeV protons (Sect. 6.1). But whenever possible, the relative gas gain as a function of the amplification field is used for a discussion of the results.

2.5 Signal Formation

The formation of signals in Micromegas has been discussed in detail in Bortfeldt (2010) and Bortfeldt et al. (2012) and is shortly summarized here. The description is inspired by a discussion of signal formation in parallel plate avalanche chambers by Mathieson and Smith (1988).

Charge signals are created by ionization processes due to traversing particles or photons followed by gas amplification and subsequent drift of ions from the anode to the mesh.

Consider a localized energy deposition in the several millimeters wide drift region e.g. by a X-ray photon with 5.9 keV, resulting in a cloud of electrons and positive ions, with a diameter on the order of 0.5 mm. Electrons and ions are separated by the electric drift field, the electrons drift with $v_d \sim 0.04$ mm/ns towards the high-field region between mesh and anode (Sect. 2.3). Depending on the dimension of the drift field, a certain fraction of electrons can enter the amplification region, the rest is lost on the micro-mesh (Sect. 2.6). Upon entering the typically 128 µm wide region between mesh and anode, the electrons trigger a Townsend avalanche, resulting in charge amplification by a factor on the order of 10^3, Sect. 2.4.

Due to the large electron mobility, the gas amplification process is finished in less than 1 ns. Charge signals start to evolve on the anode $q_a(t)$ and the mesh $q_m(t)$ as soon as electrons reach the anode, the temporal behavior of the major signal component is caused by the drift of positive ions towards the mesh. The observable signals are given by the sum of charge carriers from gas amplification on the electrode and the induction signal by charge carriers in the amplification region

$$q_a(t) = q_e(t) - q_{ai}(t)$$
$$q_m(t) = q_i(t) - q_{mi}(t) \,. \tag{2.26}$$

$q_e(t)$ is the constant charge of electrons, produced in gas amplification, $q_{ai}(t)$ is the negative mirror charge of the positive ion distribution in the amplification gap on the anode, $q_i(t)$ is the positive ion charge on the mesh and $q_{mi}(t)$ is the negative mirror charge on the mesh, produced by positive ions in the amplification region.

The four components and the resulting detectable charge signals are shown in Fig. 2.7. Details of the underlying calculation can be found in Bortfeldt (2010).

Unlike a low-energy X-ray, a charged particles creates a trace of ionization charge in the drift gap, the electrons enter the amplification region over a period on the order of 100 ns, depending on their drift velocity, diffusion and the drift gap width. Each electron creates a signal with the described temporal behavior. The resulting

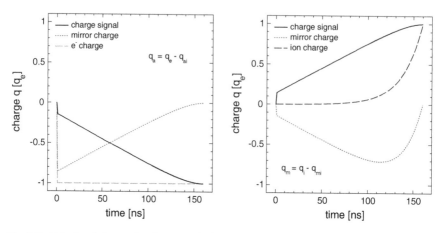

Fig. 2.7 Calculated charge signals on anode *left* and mesh *right*, assuming an amplification gap of 128 μm, an amplification field of 43.0 kV/cm and a detector gas mixture of Ar:CO$_2$ 93:7 vol% (Bortfeldt et al. 2012)

observable charge signal is determined by the ionization charge distribution, the electron drift velocity and the single electron signal evolution time, that is given by the maximum drift time of positive ions from the anode to the mesh. In order to decrease the overall signal length, the electron as well as the ion drift time has to be reduced.

The electron drift time can be controlled by a variation of the drift field, up to a certain extent. Also the use of modified Ar:CO$_2$ gas mixtures allows for a variation, see Fig. 2.4. Considerable variations however can only be achieved by using different gas admixtures such as methane or isobutane.

The ion drift velocity can only feebly be adjusted by the amplification field, since the desired gas gain defines the amplification field. The ion drift is furthermore rather insensitive to gas admixtures. In order to considerably increase the ion drift velocity, completely different gas constituents have to be used. Following Table 2.2, the use of lighter noble gases such as neon or helium allows for a considerable reduction of the signal length.

2.6 Mesh Transparency

Electrons from ionization processes in the drift region move in the homogeneous field between cathode and mesh towards the mesh. Upon entering the amplification region, avalanches are created, which can be detected by the readout electronics. The signal rise time is defined by the sum of the electron drift time in the drift region and the ion drift time in the amplification region. The electron drift time $t_d = d_d/v_d$, can reach values above several 100 ns at low fields, depending on the drift gap width d_d and the drift velocity v_d, see Fig. 2.4. If the integration time of the applied preamplifier electronics is smaller than the electron drift time, only a fraction of the ionization charge is detected, leading to smaller pulse heights at small drift fields. Furthermore, recombination with positive ions and formation of negative ions hinders the electrons from reaching the amplification region at low fields. Especially contaminations with oxygen lead to a strong attachment, due to its high electronegativity. Strictly speaking, these low-field effects are not correlated to the electron mesh transparency, although all effects, leading to the observed drift field dependence of the pulse height are often subsumed in the term mesh transparency, that is used as synonym for the transparency of the mesh for electrons.

The optical transparency of a micro-mesh with 400 lines per inch (lpi) and 18 μm wire diameter is 0.513, and only 0.368 for a wire diameter of 25 μm. Fortunately, the transparency of the mesh for electrons is much higher, as the high electric amplification field extends into the mesh holes and guides electrons around the wires into the high field region. The electron mesh transparency is strongly correlated with the transverse electron diffusion, see Fig. 2.5, since diffusion off the electric field lines leads to a loss of electrons on the mesh wires. Furthermore, the ratio of the drift and the amplification field defines the extension of the high field into the low field region. With increasing drift field, more and more field lines end on the mesh, leading to a reduction of detectable charge.

In Fig. 2.8, the electron mesh transparency for two different Ar:CO$_2$ gas mixtures and two different mesh configurations is shown. Drift of single electrons with subsequent gas amplification has been simulated with the GARFIELD++ (Veenhof 2010)

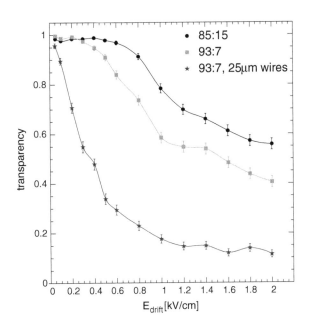

Fig. 2.8 Electron mesh transparency as a function of the drift field, calculated with GARFIELD for a woven stainless steel mesh with 63.5 μm pitch and 18 μm wire diameter for two different Ar:CO$_2$ gas mixtures at 20 °C and 1013 mbar and for a woven mesh with equal pitch, but 25 μm wire diameter. An amplification field of $E_{amp} = 39.1$ kV/cm is assumed. The simulation is described in detail in Moll (2013), results courtesy of P. Lösel

package. The simulation has been developed and validated for the investigation of discharges in Micromegas, induced by strongly ionizing particles (Moll 2013).

For low drift fields, a transparency of 0.95–0.99 is reached. Clearly visible is the influence of the increasing drift field on the electron mesh transparency. The discussed and usually observed initial increase at very low fields is not correctly simulated, as contamination through oxygen and the integration time of the preamplifier electronics are not taken into account. The thin-wire mesh has a transparency plateau close to 1 for drift fields up to $E_{drift} \sim 0.3\,kV/cm$ for the 93:7 vol% and up to $E_{drift} \sim 0.6\,kV/cm$ for the 85:15 vol% gas mixture. The difference between the two gas mixtures is caused by the different transverse electron diffusion, cf. Fig. 2.5. The observed decrease of the transparency for the thin-wire mesh is sometimes less pronounced than in the simulation, since the wire mesh is often rolled or pressed to improve electron transparency.

For the thick-wire mesh, the loss of electrons with increasing drift field is much more pronounced, as the optical transparency already is considerably smaller.

Correlated to the electron transparency is the ion mesh transparency. It is often expressed by the so called ion backflow, which denotes the fraction of ions, produced in gas amplification, that are not neutralized at the mesh, but move into the drift region between mesh and cathode. As ion backflow limits the high-rate capabilities of Micromegas with very large drift regions such as Time-Projection-Chambers or at ultra-high rates, it is usually reduced as much as possible. In Sect. 5.4, a measurement of the ion backflow ratio in a resistive strip Micromegas is presented.

2.7 Discharges

In Micromegas detectors non-destructive discharges between the mesh and the readout structure are observed. A discharge leads to a conductive plasma between the mesh and the affected part of the anode structure and causes dead time, due to the necessary recharge of anode or mesh.

A discussion of discharge development measurements in different micro-pattern gaseous detectors can be found in Bressan et al. (1999).

The probability for spontaneous discharges, caused by dust or detector imperfections, can be strongly reduced by detector assembly under clean room conditions. Furthermore in non-resistive strip Micromegas a conditioning of the detector is possible, where dust, etc. is burned in air, prior to assembly, using an elevated high-voltage between mesh and anode.

Strongly ionizing particles such as alpha particles, nuclear fragments or low-energy protons from (n,p)-reactions can create a charge density in the amplification region sufficiently large, that a mesh directed streamer develops. A critical charge density of $1.77 \times 10^6\,e/0.01\,mm^2$ has been determined by Moll (2013) in a standard Micromegas with $128\,\mu m$ amplification gap, operated with Ar:CO$_2$ 93:7 vol%.

Fig. 2.9 Streamer development in the amplification region of a Micromegas detector, following Raizer (1991)

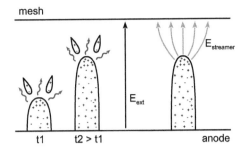

The ionization induced streamer development in a planar electrode geometry has been discussed in detail by Raether (1964) and Raizer (1991). In the following a short, qualitative summary of the mechanism is given.

Under normal operation conditions, ionization charge, created by traversing particles, drifts into the high-field region between mesh and anode, where it is amplified. Positive ions from gas amplification drift within about 150 ns to the mesh, where they are neutralized. The total charge, created by a minimum ionizing particle and subsequent amplification is typically on the order of $6 \times 10^4\, e$, too small to significantly alter the amplification field.

Strongly ionizing particles can create a much larger ion density in the amplification gap, drift of ions to the mesh competes with continuous ion production by ionization charge triggered gas avalanches. If the ion density in the proximity of the anode exceeds a critical value, the development of a mesh directed streamer begins: As the electric field of the positive ion cloud and its mirror charge on the anode is of the same order as the external amplification field, the resulting field between the upper part of the ion cloud, the so called streamer head, and the mesh is enhanced, between the ion cloud and the anode, the field is compensated. Photons from gas atom de-excitation and ion recombination leave the ion cloud and can ionize gas atoms in the vicinity. The photo-electrons trigger additional Townsend avalanches, that are directed towards the streamer head, due to the large electric field in the proximity of the streamer head, see $t = t1$ in Fig. 2.9.

The ions produced in these secondary avalanches lead to a growth of the streamer towards the mesh, $t = t2 > t1$. Due to the compensation of the external electric field between the streamer head and the anode, a quasineutral plasma with rather low conductivity is present. It leads to an equalization of potentials between the anode and the streamer. As the streamer grows towards the mesh, the electric field between mesh and streamer head increases. At a certain point secondary electron emission from the mesh sets in, either due to ion-impact ionization or photo-effect. A larger number of electrons is created, that are accelerated towards the streamer and create a well conducting quasineutral plasma, that is on mesh potential as it touches the mesh. The large electric field in front of the head of this "back-streamer" leads to considerable gas amplification, such that the back-streamer develops towards the anode, leaving behind a well conducting plasma. Once the back streamer touches the anode the mesh discharges through the plasma channel onto the anode.

This discussion shows, that discharges in Micromegas, unlike in Geiger-Müller-tubes, are localized. Their impact on the detector performance can be greatly reduced by limiting the affected anode region. This can be achieved e.g. by covering the copper anode structure with a resistive layer (Sects. 3.1.2 and 3.1.3) or by coupling copper anode strips or pixels individually to high-voltage (Sect. 3.2).

References

Allison WWM, Cobb JH (1980) Relativistic charged particle identification by energy loss. Ann Rev Nucl Part Sc 30(1):253–298. doi:10.1146/annurev.ns.30.120180.001345. http://www. annualreviews.org/doi/abs/10.1146/annurev.ns.30.120180.001345

ATLAS Collaboration (2013) New small wheel technical design report. In: Technical design report CERN-LHCC-2013-006, CERN, Geneva. http://cds.cern.ch/record/1552862

Bay A, Perroud JP, Ronga F, Derr J, Giomataris Y, Delbart A, Papadopoulos Y (2002) Study of sparking in micromegas chambers. Nucl Instrum Methods A 488(1–2):162–174. doi:10.1016/ S0168-9002(02)00510-7. http://www.sciencedirect.com/science/article/B6TJM-45BRTKJ-V/ 2/a284420c8f18198d97bd3ecbd76666d4. ISSN 0168–9002

Beringer J (Particle Data Group) et al (2012) The review of particle physics. Phys Rev D 86:010001

Bernet C, Abbon P, Ball J, Bedfer Y, Delagnes E, Giganon A, Kunne F, Le Goff J-M, Magnon A, Marchand C, Neyret D, Panebianco S, Pereira H, Platchkov S, Procureur S, Rebourgeard P, Tarte G, Thers D (2005) The 40×40 cm^2 gaseous microstrip detector micromegas for the high-luminosity COMPASS experiment at CERN. Nucl Instrum Methods A 536(1–2):61–69. doi:10.1016/j.nima.2004.07.170.   http://www.sciencedirect.com/science/article/B6TJM-4D3WCPJ-K/2/b0ba572003c1828607a8023b152aa403. ISSN 0168-9002

Biagi SF (1999) Monte Carlo simulation of electron drift and diffusion in counting gases under the influence of electric and magnetic fields. Nucl Instrum Methods A 421(1–2): 234–240. doi:10.1016/S0168-9002(98)01233-9. http://www.sciencedirect.com/science/article/ pii/S0168900298012339. ISSN 0168–9002

Bichsel H (1988) Straggling in thin silicon detectors. Rev Mod Phys 60:663–699. doi:10.1103/ RevModPhys.60.663. http://link.aps.org/doi/10.1103/RevModPhys.60.663

Bichsel H (2006) A method to improve tracking and particle identification in TPCs and silicon detectors. Nucl Instrum Methods A 562(1):154–197. doi:10.1016/j. nima.2006.03.009.   http://www.sciencedirect.com/science/article/B6TJM-4JNFN6M-3/2/ 5c93a031604810dc6d2902efef4044fe. ISSN 0168–9002

Bortfeldt J (2010) Development of micro-pattern gaseous detectors—micromegas. Diploma thesis, Ludwig-Maximilians-Universität München, http://www.etp.physik.uni-muenchen.de/ dokumente/thesis/dipl_bortfeldt.pdf

Bortfeldt J, Biebel O, Heereman D, Hertenberger R (2012) Development of a high-resolution muon tracking system based on micropattern detectors. IEEE Trans Nucl Sci 59(4):1252–1258. doi:10.1109/TNS.2012.2202688. ISSN 0018–9499

Bressan A, Hoch M, Pagano P, Ropelewski L, Sauli F, Biagi S, Buzulutskov A, Gruwé M, De Lentdecker G, Moermann D, Sharma A (1999) High rate behavior and discharge limits in micro-pattern detectors. Nucl Instrum Methods A 424(23):321–342. doi:10.1016/S0168-9002(98)01317-5http://www.sciencedirect.com/science/article/pii/S0168900298013175. ISSN 0168–9002

Čerenkov PA (1937) Visible radiation produced by electrons moving in a medium with velocities exceeding that of light. Phys Rev 52:378–379. doi:10.1103/PhysRev.52.378. http://link.aps.org/ doi/10.1103/PhysRev.52.378

Deslattes RD, Kessler EG Jr., Indelicato P, de Billy L, Lindroth E, Anton J, Coursey JS, Schwab DJ, Chang C, Sukumar R, Olsen K, Dragoset RA (2005) X-ray transition energies (version 1.2). National Institute of Standards and Technology, Gaithersburg. http://physics.nist.gov/XrayTrans

Grupen C, Shwartz B (2008) Particle Detectors. Cambridge University Press, Cambridge

Gupta M (2013) Calculation of radiation length in materials. PH-EP-Tech-Note-2010-013

Ipe NE, Fassò A (1994) Gas bremsstrahlung considerations in the shielding design of the advanced photon source synchrotron radiation beam lines. Nucl Instrum Methods A 351(2–3): 534–544. doi:10.1016/0168-9002(94)91383-8http://www.sciencedirect.com/science/article/pii/0168900294913838. ISSN 0168–9002

Kleinknecht K (1992) Detektoren für Teilchenstrahlung. Teubner

Kuger F (2013) Simulationsstudien und Messungen zu Gasverstärkungsprozessen in Micromegas für den Einsatz im ATLAS NewSmallWheel. Master's thesis, Julius-Maximilians-Universität Würzburg, Germany

Lippert B (2012) Studien zur Signalentstehung und Parametrisierung der Gasverstärkung in einem Micromegas-Detektor. Bachelor's thesis, Ludwig-Maximilians-Universität München. unpublished, day-to-day supervision by J. Bortfeldt

MacDaniel EW, Mason EA (1973) The Mobility and Diffusion of Ions in Gases. Wiley, New York

Mathieson E, Smith GC (1988) Charge distributions in parallel plate avalanche chambers. Nucl Instrum Methods A 273(2–3):518–521. doi:10.1016/0168-9002(88)90046-0http://www.sciencedirect.com/science/article/B6TJM-473DCY1-B6/2/69b865447a3e9dfb74709952ea408006. ISSN 0168–9002

Moll S (2013) Entladungsstudien an Micromegas-Teilchendetektoren. Diploma thesis, Ludwig-Maximilians-Universität München, Germany

Raether H (1964) Electron avalanches and breakdown in gases. Butterworths, London

Raizer YP (1991) Gas discharge physics. Springer, New York

Şahin Ö, Tapan I, Özmutlu EN, Veenhof R (2010) Penning transfer in argon-based gas mixtures. JINST 5(05):P05002 http://stacks.iop.org/1748-0221/5/i=05/a=P05002

Sternheimer RM, Berger MJ, Seltzer SM (1984) Density effect for the ionization loss of charged particles in various substances. Atom Data Nucl Data 30(2):261–271. doi:10.1016/0092-640X(84)90002-0http://www.sciencedirect.com/science/article/pii/0092640X84900020. ISSN 0092–640X

Thibaud F, Abbon P, Andrieux V, Anfreville M, Bedfer Y, Burtin E, Capozza L, Coquelet C, Curiel Q, d'Hose N, Desforge D, Dupraz K, Durand R, Ferrero A, Giganon A, Jourde D, Kunne F, Magnon A, Makke N, Marchand C, Neyret D, Paul B, Platchkov S, Usseglio M, Vandenbroucke M (2014) Performance of large pixelised micromegas detectors in the COMPASS environment. JINST 9(02):C02005 http://stacks.iop.org/1748-0221/9/i=02/a=C02005

Townsend JS (1910) The Theory of Ionization of Gases By Collision. Constable and Company Ltd, London

Veenhof R (2010) Garfield—simulation of gaseous detectors. http://garfield.web.cern.ch/garfield/

Chapter 3
Floating Strip Micromegas

3.1 Current Micromegas Types

In the following section, currently used Micromegas designs are presented and shortly discussed. The internal setup and performance of the original Micromegas detector with directly read out copper anode strips, is summarized. In order to reduce the effect of discharges between mesh and anode, resistive strip Micromegas have been developed, in which the copper anode strips are covered with a high-ohmic, micro-structured material (Alexopoulos et al. 2011). The resistive anode and the copper readout strips are separated by a thin sheet of insulator material. As a third type, an integrated Micromegas, built by silicon waver post-processing directly on pixel CMOS chips like the Timepix is introduced (Chefdeville et al. 2006). The three Micromegas types differ in the specific realization of the amplification structure, the drift region is unchanged.

3.1.1 Standard Micromegas

In standard Micromegas, the amplification gap is formed by a thin woven or electro-formed micro-mesh, held by supportive pillars at a typical distance of 128 µm to the readout structure, which consists of bare or Ni-Au-covered copper strips. The internal setup of a standard Micromegas is schematically shown in Fig. 3.1.

Since the photo-lithographic production of copper strips is a conventional process, used in industrial printed circuit board production, readout structures for standard Micromegas are relatively easy to produce and furthermore cheap. Since the amount of non-metal material in the detector can be limited to a minimum, no aging due to polymerization of detector components is expected. Due to the very good conductivity of copper, no charge-up of the readout structure, besides the small gaps between neighboring strips, is expected.

© Springer International Publishing Switzerland 2015
J. Bortfeldt, *The Floating Strip Micromegas Detector*, Springer Theses,
DOI 10.1007/978-3-319-18893-5_3

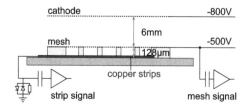

Fig. 3.1 Schematic view of a standard Micromegas detector. The anode strips are usually read out with highly-integrated electronics, such as the Gassiplex or the APV25 chip (Sects. 4.1.1 and 4.1.2). The inverse signal on the mesh, decoupled over a capacitor, can e.g. be used for triggering purposes

For a summary of the pulse height behavior, efficiency and spatial resolution achievable with standard Micromegas see Sect. 5.1. In typical applications, gas gains on the order of 5000 are used, the detection efficiency for minimum ionizing particles is above 98 % and a spatial resolution on the order of 35–50 μm can be reached.

As discussed in Sect. 2.7, instantaneous charge densities in the amplification gap exceeding approximately $2 \times 10^8 \ e/\mathrm{mm}^2$, lead to non-destructive discharges between mesh and anode copper strips. The readout electronics has to be protected against large discharge currents by a protective circuit, connected to each readout strips. Due to this circuit and the good conductivity of the discharge plasma channel, the mesh high-voltage drops to zero in a discharge. During the recharge time, the whole detector is insensitive to traversing particles. Due to the sufficiently large heat capacitance and conductivity of 35 μm thick copper anode strips, discharges seem not to damage the readout structure.

The dead time after a discharge can be calculated from the recharge behavior of a capacitor

$$U(t) = U_0 \left(1 - \exp\left(-\frac{t}{RC}\right)\right) , \qquad (3.1)$$

where C is the overall mesh-to-ground capacitance and R the used mesh recharge resistor. Assuming, that the detector is efficient, as soon as the actual amplification field deviates by less than 1 % from the nominal value, the dead time is given by

$$t_{\mathrm{dead}} \sim 5RC . \qquad (3.2)$$

The dead time of a $9 \times 10 \, \mathrm{cm}^2$ standard Micromegas with $C \sim 1.5 \, \mathrm{nF}$ and $R = 10 \, \mathrm{M}\Omega$ is approximately $5RC = 75 \, \mathrm{ms}$.

In applications with discharge rates below 1 Hz, this efficiency drop can be tolerated. In high-rate minimum ionizing particle detection application with elevated neutron and photon background, alternative amplification structures incorporating resistive strips or floating strips are recommended. Discharge probabilities range from 10^{-4} in high-energy pion beams to 10^{-7} in low energy ion beams.

3.1.2 Resistive Strip Micromegas

In resistive strip Micromegas, the amplification region with a typical width of 128 μm is formed by a micro-mesh and a layer of resistive strips. The resistive strips usually consist of a mixture of carbon and phenolic resins (Michael GmbH 2013), where the resistivity can be controlled by the amount of added carbon powder. By applying high-voltage to the resistive strips and grounding the mesh or vice versa, typical amplification fields of 38 kV/cm are created. A schematic cut through a resistive strip Micromegas with one-dimensional strip readout is shown in Fig. 3.2.

Signals couple capacitively from the resistive strips to a layer of copper anode strips, separated by a thin layer of insulating prepreg FR4-like material: Electrons, created in gas avalanches, are collected locally on resistive strips and are released as the positive ions drift towards the mesh. The electron charge then spreads and flows along the resistive strip to the high-voltage supply.

In a discharge between mesh and resistive strips, the involved resistive strips charge up only locally, such that the electric field quickly reduces to zero, interrupting the discharge. Due to the small extension of the affected area, the involved detector capacitance is very small, such that the overall voltage drop is only on the order of a few 10 mV, which does not affect the detection efficiency of the Micromegas outside the affected region.

Resistive strip Micromegas are very well suited for applications with medium hit rates up to several $100 \, \text{kHz/cm}^2$ and with typical discharge rates on the order of several Hz due to their superior discharge suppression capabilities. Similar gas gains, efficiencies and spatial resolutions as with standard Micromegas can be reached. Due to the capacitive coupling of resistive strips to the readout structure, a two dimensional readout with two stacked perpendicular layers of copper readout strips can be easily realized (Byszewski and Wotschack 2012). By correctly adapting the width of strips in the layer, closer to the resistive strips, similar pulse height on both readout layers can be reached. Micromegas with three layers of readout strips, relatively rotated by angles of $60°$, have been realized, allowing for reliably resolving hit ambiguities.

Due to the necessity of amplification charge drain over the complete length of the resistive strips, a rate-dependent charge-up in resistive strip Micromegas is observed.

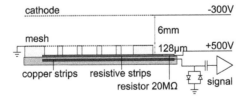

Fig. 3.2 Schematic view of a resistive strip Micromegas detector. Signals couple capacitively to copper readout strips, connected to the readout electronics. Resistive strips have a resistivity of $0.5–100 \, \text{M}\Omega/\text{cm}$, each resistive strip is individually connected to high-voltage via a printed resistor with resistivity of $10–200 \, \text{M}\Omega$

Furthermore is the construction process of resistive strip Micromegas complicated, especially due to the deposition of resistive material. The production of the resistive strips with the necessary relative precision can either be achieved by filling photo-lithographically etched masks of solder-resist with resistive material or by directly screen-printing the resistive strips.

An alternative method has been proposed by (Ochi et al. 2013) and is currently under development: A 25 μm thick Kapton foil is coated with photo resist, the desired resistive strip pattern is photo-lithographically etched into the photo resist. A thin tungsten layer is homogeneously sputtered onto the foil, followed by a thin sputtered layer of carbon. The photo resist is then washed away, yielding a finely structured layer of mechanically stable resistive strips. With a 1 nm thick tungsten layer and a 30 nm thick carbon layer, a resistivity of $10\,\text{M}\Omega/\square$ can be reached. It can be adjusted over the respective layer thickness. The Kapton foils with the resistive structure are then glued to a printed circuit board, carrying the copper readout structure. This method has the large advantage of separating readout and resistive structure produc-tion, furthermore very fine patterns can be reached.

The aging behavior of resistive strip Micromegas is currently under investigation. Due to the presence of plastic based material in the active volume, an increased aging probability might be present. In the literature published up to now (Galán et al. 2013; Jeanneau et al. 2012), no aging after irradiation with thermal neutrons and X-rays has been observed though.

3.1.3 InGrid—Integrated Silicon Waver Based Micromegas

The combination of a CMOS pixel readout chip, originally intended for charge sig-nal acquisition in semiconductor X-ray converters, and a Micromegas amplification structure, has been proposed by Campbell et al. (2005). The applied Medipix2 chip (e.g. San Segundo Bello et al. 2003) features 256×256 pixels with dimensions $55 \times 55\,\mu\text{m}^2$.

Production of the Micromegas structure by silicon waver post-processing has been proposed by Chefdeville et al. (2006), considerably simplifying the produc-tion process and improving the performance. The device has been named InGrid.[1] Mesh-supporting pillars consist of photo-lithographically etched photo-resist, the micro-mesh is formed out of a 0.8 μm thick aluminum layer.

InGrid detectors work according to the same principles as standard Micromegas (Chap. 2). Due to the fine granulation of the readout structure, the detection of single electrons, produced by a traversing charged particle or X-ray, is possible, δ-rays can be resolved. The amplification region, confined by the readout CMOS chip and a thin aluminum mesh, is precisely defined, resulting in a very good energy resolution of 12.3 % FWHM for 5.9 keV X-rays (Chefdeville et al. 2008).

---

[1] INtegrated GRID.

The deposition of a silicon based resistive layer considerably reduces the damage to the sensitive device due to discharges (Bilevych et al. 2011).

Due to their superior resolution, InGrid detectors are e.g. proposed for X-ray detection in low background experiments (Krieger et al. 2013).

3.2 The Floating Strip Principle

In the following section, the underlying ideas of the floating strip Micromegas principle are introduced and discussed. Different possible realizations are presented.

3.2.1 The Idea

Inspired by the discharge suppression mechanism in resistive strip Micromegas, but aiming at avoiding the use of plastic and resistive material in the active volume, the floating strip principle has been developed. Its functional principle is demonstrated in Fig. 3.3. The drift region is unchanged, the amplification region is formed by a micro-mesh and copper anode strips, that are individually connected to high-voltage via large resistors with $R \gtrsim 20\,\mathrm{M\Omega}$. Signals on the anode strips are decoupled by small high-voltage resistant capacitors, their capacitance is of the same order as the anode-strip-to-mesh- and strip-to-strip capacitance. This leads to a loss of signal but enables a powerful discharge suppression.

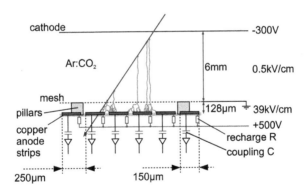

Fig. 3.3 Functional principle of a floating strip Micromegas. A charged particle (*dark blue line*) ionizes the gas in the active region of the detector. The ionization electrons drift into the high-field region between mesh and anode strips, where they are amplified in Townsend avalanches. The anode strips are individually connected to high-voltage via resistors (above $20\,\mathrm{M\Omega}$), signals are decoupled using capacitors and amplified by fast charge sensitive preamplifiers

A similar powering scheme has been proposed by Bay et al. (2002), but avoiding the extreme recharge resistor and coupling capacitor values, which ultimately enable the discharge suppression.

The ionization, gas amplification and signal formation processes in floating strip Micromegas are equal to those in standard Micromegas, see Chap. 2. In discharges however, floating strip Micromegas behave differently: In a discharge, the potential of the affected strips can "float", such that the potential difference between mesh and strips quickly levels, disrupting the discharge. The other strips remain unaffected and thus the detector is fully efficient, except for the few strips, participating in the discharge. Due to the small overall strip capacitance on the order of 1 pF/cm, the recharge time of individual strips after a discharge of approximately 1 ms is short.

Comparing the dead time of a $9 \times 10\,\mathrm{cm}^2$ standard Micromegas with 360 strips to a floating strip Micromegas of equal size, shows the advantage of the floating strip principle: Assuming an overall mesh-to-ground capacitance of 1.5 nF for the standard Micromegas, where the mesh is recharged via a 10 MΩ resistor, we expect a dead time after a discharge of 75 ms, see Eq. (3.2). Considering a comparable floating strip detector with individual strip recharge resistors of 22 MΩ and strip coupling capacitors of 10 pF i.e. an overall strip capacitance of $1.5\,\mathrm{nF}/360 + 10\,\mathrm{pF} = 14\,\mathrm{pF}$, we expect a single strip dead time of 1.5 ms. We can furthermore assume (Sect. 3.3.4) that a discharge affects 6 strips in the floating strip detector.

Using Poisson statistics and assuming a mean discharge rate of $f_{\mathrm{discharge}} = 5\,\mathrm{Hz}$, the inefficiency $\not\epsilon$ of the detector is given by

$$\not\epsilon = P(k \geq 1)c = \left(1 - \frac{\lambda^0}{0!}e^{-\lambda}\right)c\,, \tag{3.3}$$

where c is the fraction of the active area, that is affected by a discharge and $\lambda = t_{\mathrm{dead}} f_{\mathrm{discharge}}$ is given by the dead time and the discharge rate. Note, that the inefficiency is thus slightly overestimated, as a second discharge cannot be triggered in a still inefficient detector region.

For the standard Micromegas with $c = 1$, the inefficiency is $\not\epsilon = 0.31$ for $\lambda = 0.375$. For the floating strip detector with $c = 6/360$ and $\lambda = 7.5 \times 10^{-3}$ on the other hand, the expected inefficiency is only $\not\epsilon = 1.3 \times 10^{-4}$ and thus more than three orders of magnitude lower.

3.2.2 Realization—Discrete and Integrated Floating Strip Micromegas

In floating strip Micromegas, anode strips are individually connected to high-voltage over high-ohmic resistor, signals are decoupled over small capacitors. In the course of this thesis, two different realizations of floating strip Micromegas are constructed:

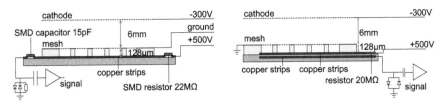

Fig. 3.4 Two possible realizations of floating strip Micromegas. Using discrete SMD components (*left*) allows for an optimization of the used components due to easy exchangeability. Higher anode strip pitch and less tedious construction are the advantages of the integrated solution (*right*)

First, with discrete SMD[2] capacitors and resistors, and second, with integrated capacitors and printed resistors, Fig. 3.4.

The use of exchangeable SMD components allows for an optimization of the chosen values and furthermore enables the direct comparison of standard and floating strip Micromegas, as the behavior of a standard Micromegas can be simulated by choosing small recharge resistors. Due to the necessary high-voltage sustainability, the SMD component package must not be smaller than 0805, corresponding to foot print dimensions of $2.0 \times 1.2\,mm^2$. This limits the minimum achievable strip pitch to about 0.5 mm. Some dielectrics used in SMD capacitors, change their permittivity considerably in the presence of a large DC voltage, leading to a considerable decrease of the capacitance, see e.g. Sect. 5.3.6. The measurable pulse height can be significantly reduced.

For larger or already optimized floating strip Micromegas, the integrated type is advantageous. The individual strip resistors are replaced by printed resistors, consisting of a mixture of carbon and phenolic resin, equal to the resistive strip material. Intrinsically high-voltage resistant capacitors are formed by adding a second layer of readout strips below the anode strips, separated by a thin layer of insulating FR4. Conventional multi-layer printed circuit boards or the use of double-clad thin FR4 material are equally suitable. The production process of integrated floating strip Micromegas is less tedious due to the avoidance of SMD component soldering, at the price of not being able to adapt resistances and capacitances afterwards.

3.3 Discharge Performance

The advantage of floating strip over standard Micromegas is seeded in the significantly improved discharge behavior. In the following, direct measurements of the behavior of a floating strip Micromegas under discharges are discussed. The detector has been modeled with the circuit simulation program LTSpice (Linear Technology Corporation 2010), overall voltage drop and the microscopic structure and processes within discharges have been measured and quantitatively compared to the simulation.

---

[2] Surface-Mounted Device.

3.3.1 Experimental Investigation of Discharges

The discharge behavior of a small floating strip Micromegas has been experimentally investigated. Its $6.4 \times 6.4\,cm^2$ active area is formed by 128 copper anode strips with $300\,\mu m$ width and $500\,\mu m$ pitch, individually coupled to high-voltage via exchangeable SMD resistors and exchangeable 15 pF SMD capacitors. Discharges were induced by strongly ionizing alpha particles with $E_\alpha = 5.2\text{--}5.8\,MeV$ from a mixed nuclide ^{244}Cm-^{241}Am-^{239}Pu-source.

The setup is shown schematically in Fig. 3.5. The direct voltage signals from seven neighboring strips were, after attenuation with a $1000\,k\Omega{:}1\,k\Omega$ voltage divider, recorded with two 4 channel Tektronix MSO 4104 oscilloscopes (Tektronix Inc. 2013). Discharges were detected, by monitoring the common strip high-voltage over a 10 nF capacitor with the remaining oscilloscope channel. Upon detection of a discharge, producing an internal trigger signal, the other oscilloscope was likewise triggered and raw data from all eight channels were sent via Ethernet to a DAQ computer. A custom LabVIEW[3] program has been written to control data acquisition and recording.

A typical discharge signal is shown in Fig. 3.6. The direct discharge signal is visible on strips 1, 2 and 3, and is capacitively coupled to the neighboring strips. With increasing distance from the discharge, the pulse height of the capacitively coupled signal decays quickly. Note that the vertical signal offset is due to the oscilloscope acquisition window, the baseline before the discharge is actually at 0 V. Due to the applied voltage divider, the measured strip signals are by a factor of 1000 smaller, than the true strip signals, the global voltage drop on the other hand is measured directly. As a first result we can conclude, that the global voltage drop after a discharge in floating strip Micromegas, which affects all strips, is below 1 V.

The acquired strip signals are fitted with a piece-wise defined, five-parameter function

Fig. 3.5 Schematic setup used in the discharge behavior investigation in a small floating strip Micromegas detector. Eight oscilloscope channels were used to record the voltage signal on seven neighboring strips and on the global high-voltage distributor

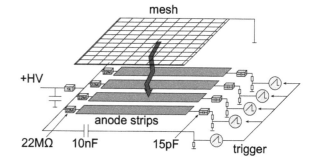

---

[3]http://www.ni.com/labview/.

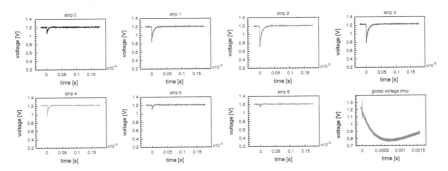

Fig. 3.6 Typical signal for a discharge onto strips 1, 2, and 3. The offset of all signals is an artifact of the oscilloscope acquisition window. The small strip coupling capacitor and the voltage divider form a high pass filter, such that the acquired strip signals are shorter then the true signals. The temporal behavior of the global voltage drop is not distorted

$$
f(t) = \begin{cases} b & \text{for } t < t_0 \\ -\frac{p}{t_1 - t_0}(t - t_0) + b & \text{for } t_0 \leq t < t_1 \\ b - p \exp\left(-\frac{t - t_1}{t_d}\right) & \text{for } t \geq t_1 \, . \end{cases} \tag{3.4}
$$

The offset of the signal is given by b, the discharge signals begins at t_0 and reaches its minimum at t_1. The pulse height of the signal is given by p and the signal decay time is described by t_d.

The absolute global voltage drop is given by the difference between the measured signal offset and the minimum value.

The measured voltage drops and strip signals are discussed in detail in Sects. 3.3.3 and 3.3.4.

3.3.2 LTSpice Detector Model

In order to understand the relation between the discharge performance and the detector geometry, used coupling capacitors and high-voltage resistor, discharges in a floating strip Micromegas have been simulated with the circuit simulation program LTSpice. The detector equivalent circuit, the high-voltage supply and the readout electronics, assumed in the simulation, is shown in Fig. 3.7.

An explanation of the used symbol names and their values are given in Table. 3.1. Discharges between the mesh and single anode strips are modeled as low-ohmic connection with a duration of 10 ns. This agrees with the mesh-directed streamer model for discharges, discussed in Sect. 2.7. Furthermore space charge induced, conductive connections between neighboring strips are simulated, see Sect. 3.3.4 for an explanation. The voltage signal on eight neighboring strips is directly read out over a 1000:1 voltage divider. This agrees with the realized experimental setup. The

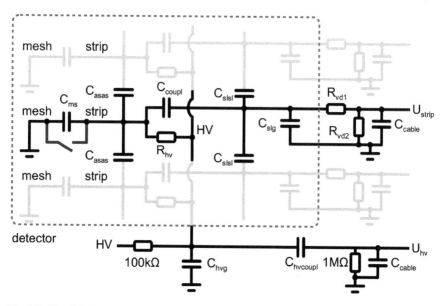

Fig. 3.7 Simplified schematic of the floating strip Micromegas, used in the discharge measurements. In a LTSpice simulation, eight neighboring strips have been fully simulated. A discharge is modeled as low-ohmic connection between mesh and single anode strips (*blue* switch)

Table 3.1 Explanation of symbol names of the constituents of the LTSpice detector model, Fig. 3.7

Name	Value	Description
C_{ms}	4.77 pF	Capacitance between mesh and anode strip, incorporating anode strip to ground capacitance
C_{asas}	4.86 pF	Capacitance between neighboring anode strips
C_{coupl}	3 pF	Coupling capacitance between an anode and the corresponding readout stripline
R_{hv}	22 MΩ	Individual high-voltage recharge resistor
C_{slsl}	0.8 pF	Capacitance between neighboring striplines
C_{slg}	0.2 pF	Capacitance between each strip line and ground
R_{vd1}	1 MΩ	First resistor of signal voltage divider
R_{vd2}	1 kΩ	Second resistor of signal voltage divider
C_{cable}	40 pF	Capacitance of the 50 Ω cable between the detector and the used oscilloscopes
C_{hvg}	20 nF	Buffer capacitor in the high-voltage filter
$C_{hvcoupl}$	10 nF	Coupling capacitor, used to measure the high-voltage drop

The values are calculated for the $6.4 \times 6.4 \, cm^2$ floating strip Micromegas, that has been used in the discharge measurements

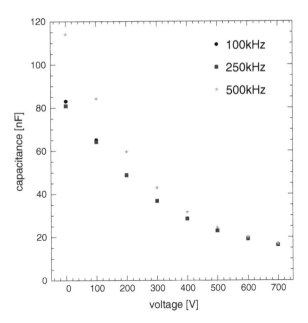

Fig. 3.8 Measured capacitance of the 100 nF high-voltage buffer capacitor as a function of the applied DC voltage

high-voltage, provided by the HV supply, is filtered with a low pass, global voltage drops are monitored over a capacitor.

In the detector equivalent circuit, only capacitances and explicit ohmic resistors are considered. All capacitances are either calculated from geometric considerations or directly measured. The mesh-to-anode-strip and the stripline-to-ground capacitances C_{ms} and C_{slg} are calculated using the formula for the capacitance of a parallel plate capacitor. The formula, used to calculate the capacitance between neighboring anode strips and neighboring striplines C_{asas} and C_{slsl}, has been taken from Nührmann (2002).

The capacitance and inductance of the cable,[4] used to supply high-voltage to the anode strips, is considered in the simulation, although it is not shown in Fig. 3.7. As the capacitor in the high-voltage filter considerably changes its capacitance as a function of the applied voltage, Fig. 3.8, the measured value of 20 nF has been used instead of the nominal value. Since the anode strip coupling capacitors C_{coupl} consists of the same dielectric, an equal capacitance reduction by 80 % from the nominal value of 15 pF is assumed.

The model has two free parameters: First, the response of the high-voltage supply to a discharge is modeled by an additional low pass filter with 2 Hz cutoff frequency. Second, the conductive connection between two neighboring strips (see Sect. 3.3.4) is assumed to cease as soon as their potential difference drops below 220 V.

The simulation results are presented and compared to measurement results in Sect. 3.3.4.

---

[4]Type: RG 58 CU, distributed capacitance and inductance $c = 100$ pF/m and $l = 250$ nH/m.

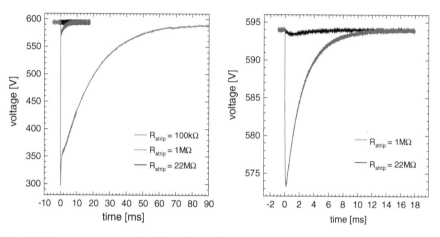

Fig. 3.9 Average common strip voltage after a discharge for three different recharge resistor configurations (*left*), measured during detector commissioning. On the right, a zoom into the left plot is shown, to demonstrate the considerable reduction of the overall voltage drop

3.3.3 Average Voltage Drop After Discharges

The global voltage drop i.e. the behavior of the common strip high-voltage after a discharge has been measured for different resistor configurations during the commissioning process.[5] The behavior of a standard Micromegas could be approximated by choosing relatively low strip resistances of $100\,k\Omega$ and including a $10\,M\Omega$ series resistor in the common high-voltage line. As a first optimization step of the resistivity configuration for the floating strip detector, the strip resistance has been increased to $1\,M\Omega$, the series resistor was adapted to $2.4\,M\Omega$. Its value has been found in a simplified LTSpice simulation with the requirement, that the high-voltage recharge current is equal to the previous configuration. For the final configuration, a strip resistor with $22\,M\Omega$ has been chosen, discarding the global series resistor. The strip resistance is a compromise between fast discharge quenching and small strip recharge time after a discharge.

In Fig. 3.9 the measured average signal on the common strip high-voltage line after a discharge is shown for the three different configurations. For the lowest strip resistivity, the global high-voltage breaks down to about 50 % of the nominal value, which is again reached by recharging after about 80 ms. During this time, the detector is inefficient. Choosing a medium strip resistivity, already greatly improves the behavior. A mean drop of only 21 V is observed, the nominal high voltage is reached again after 10 ms.

In the optimized strip configuration ($R_{\text{strip}} = 22\,M\Omega$), discharges lead to a negligible overall voltage drop of only 0.5 V. These measurements show directly, that the floating strip principle works and that it has a very positive influence on the dis-

---

[5]Excerpts from this section have been published by Bortfeldt et al. (2013).

charge performance of Micromegas detectors. In the next section, the behavior of the common strip high-voltage and the signals on individual strips in discharges is investigated in more detail.

3.3.4 Microscopic Structures of Discharges

The discharge behavior of a small optimized floating strip Micromegas has been investigated, using the experimental setup, described in Sect. 3.3.1. The signals on seven neighboring strips and on the common high-voltage line, created by alpha particle induced discharges, were recorded with two oscilloscopes. A detailed LTSpice detector simulation (Sect. 3.3.2) is used for investigating the different sub-processes in a discharge.

Around 10^4 recorded discharge signals are used for the analysis, no degradation of the detector performance has been observed under the measurements.

The distribution of measured global high-voltage drops is shown in Fig. 3.10. The global voltage drop is obviously quantized, three distinct peaks are visible. The peak centers are at $U_1 = 345$ mV, $U_2 = 445$ mV and $U_3 = 550$ mV. Similar measurements were performed by Bay et al. (2002). The peaks correspond to discharges onto one, two and three strips as explained in the following.

Discharges also create signals on strips. The pulse height depends on whether the strip participates in the discharge and on the distance to the discharge. The discharge pulse height spectrum of single strips has a complicated sub-structure with many discrete peaks. Investigating the correlation between the pulse height of different strips, allows for unfolding the complicated spectra. In this discussion, we use the pulse height correlation between strips 2 and 3 for arbitrary discharge position. In principle, correlation from any set of neighboring or close-by strips can be used, the choice is somewhat arbitrary but without loss of generality.

Discharges are separated into three different classes, depending on whether their global voltage drop corresponds to the first, second or third peak in Fig. 3.10. For all discharges in each class, the pulse height of strip 2 is drawn as a function of the pulse height of strip 3, Figs. 3.11, 3.12 and 3.13. Several clearly separated regions are visible, that correspond to different discharge processes, the observed symmetry is caused by the periodic strip structure of the readout plane.

Superimposed are black circles, representing the pulse height correlations, simulated with LTSpice. The simulation result defines the center of each circle, its radius is a measure for the uncertainty of the simulated value. A good quantitative agreement between the measured and the simulated spots is visible, the simulation is used to identify the six different discharge processes for these three classes:

- In discharges involving one strip (Fig. 3.11), the strip is discharged completely onto the mesh. This leads to a large electric field between the central and the two neighboring strips.

Fig. 3.10 Measured global voltage drop, the colored lines correspond to simulated voltage drops for discharges involving one (*red, fine dashed*), two (*green, dash-dotted*) and three strips (*blue, dashed*)

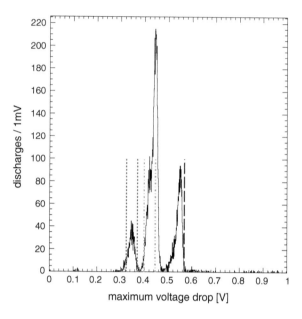

Fig. 3.11 Pulse height on strip 2 versus pulse height on strip 3, global voltage drop corresponds to the lowest peak in Fig. 3.10 i.e. discharges on a single strip

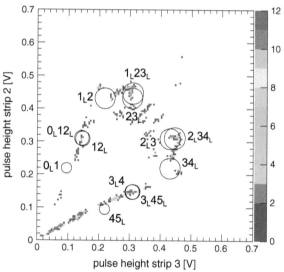

1. Enhanced by the later arriving ionization charge, charge can leak from the central strip onto the two neighboring strips after the discharge has ceased (marked with e.g. $2_L 34_L$, this denotes a discharge onto strip 3 with subsequent charge leakage onto strips 2 and 4). The subscript L denotes the strip, onto which charge has leaked.

Fig. 3.12 Pulse height on strip 2 versus pulse height on strip 3, global voltage drop corresponds to the middle peak in Fig. 3.10 i.e. discharges involving two strips

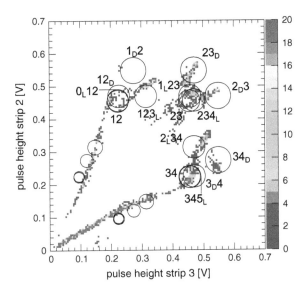

Fig. 3.13 Pulse height on strip 2 versus pulse height on strip 3, global voltage drop corresponds to the highest peak in Fig. 3.10 i.e. discharges involving three strips

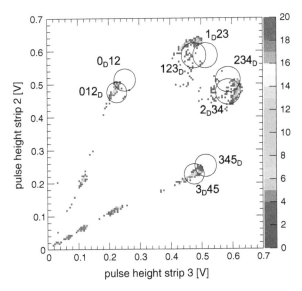

2. There are some events, in which charge leaks only onto one of the two neighboring strips (denoted with e.g. 34_L).

It is assumed, that the conductive connection between two strips is interrupted, when their potential difference decreases below 220 V.

• For discharges onto two strips (Fig. 3.12), three sub-processes exist.

3. Simultaneous discharges onto two strips (marked with e.g. 23), create equally high signals on both strips.
4. Due to the inclination of the triggering alpha particle's track, the discharge onto the second strip can occur after the conductive connection between the mesh and the first strip is already interrupted (named e.g. 23_D). Strips involved in these delayed discharges are marked with the subscript D.
5. Charge can leak onto a neighboring strip, after the simultaneous discharge on two strips has ceased (denoted with e.g. 234_L)

- Only a single process is observed for discharges onto three strips (Fig. 3.12).

6. After the contemporaneous discharge between two strips and the mesh terminated, a third strip discharges onto the mesh (named e.g. 234_D). Charge leakage onto neighboring strips is not observed, this is due to the small size of the ionization charge distribution.

If the track of the incoming alpha particle is perpendicular to the strip plane, a discharge on only one or two strips is triggered. If the inclination angle becomes too large, the critical charge density, necessary to trigger streamer development is not reached anymore.

3.3.5 Conclusions

The discharge behavior of a small floating strip Micromegas has been investigated in depth with alpha particle induced discharge experiments and a detailed LTSpice detector simulation. Two major properties could be shown:

First, the discharges in floating strip detectors are localized, depending on readout strip pitch, the nature of the strongly ionizing particle and the gas gain, a small number of anode strips is affected. In the presented measurements, discharges involving up to three neighboring strips were observed. The discharge does not spread, once triggered, across the whole detector.

Second, the common strip high-voltage drops, depending on the number of strips involved in a discharge, by a few 100 mV. This drop does not significantly affect the performance of the Micromegas detector and can be quickly restored.

Both results show the good discharge sustainability of floating strip Micromegas and suggest further improvements. A floating pixel Micromegas, in which pixels, forming the highly segmented pixelized readout structure, are individually connected to high-voltage, would suffer even less from discharge induced efficiency degrading.

3.4 Description of the Constructed Floating Strip Micromegas

Three different floating strip Micromegas have been constructed for the measurements, presented in this thesis. Their internal setup is described in the following. A small floating strip detector with discrete capacitors and resistors (Sect. 3.4.1) and a small semi-integrated detector doublet with very small material budget (Sect. 3.4.2) have entirely been built at the LMU Munich. The readout structure for a large integrated floating strip Micromegas (Sect. 3.4.3) has been constructed and built at CERN, the remaining detector components have been constructed at LMU. All detectors were assembled and commissioned by the author.

3.4.1 6.4 × 6.4 cm² Detector with Discrete Capacitors

A small floating strip Micromegas, with discrete capacitors and resistors has been used for the discharge measurements, presented in Sect. 3.3. A second detector with identical readout structure was used in cosmic muon measurements under proton irradiation, Sect. 5.3.

The internal setup of the detector, used in the discharge measurements, is shown schematically in Fig. 3.14.

Its $6.4 \times 6.4 \, cm^2$ active area is formed by 128 copper anode strips, with a pitch of 500 μm and a width of 300 μm. The 35 μm thick copper readout strips have been photolithographically etched, the lower side of the 1.5 mm thick printed circuit board, that carries the anode strips, is covered by a solid copper plane for shielding reasons. Two layers of 75 μm thick photostructurable solder resist film form the 150 μm high mesh supporting pillars and insulate the anode strips from the aluminum gas frame.

The amplification gap is closed by a 400 lines per inch woven stainless steel mesh with 25 μm thick wires, that is glued to the 6 mm thick aluminum gas frame.

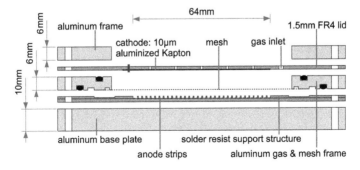

Fig. 3.14 Exploded view of a floating strip Micromegas with active area of $6.4 \times 6.4 \, cm^2$, built with discrete coupling capacitors and recharge resistors

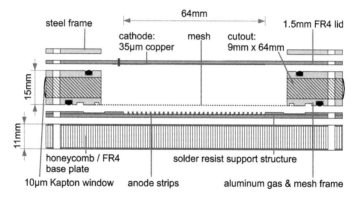

Fig. 3.15 Exploded view of a floating strip Micromegas with active area of $6.4 \times 6.4\,cm^2$, built with discrete coupling capacitors and recharge resistors. The detector has been used for cosmic muon tracking under proton irradiation

A 1.5 mm thick copper-clad FR4 board is used as detector lid. 1 mm holes are drilled into the lid, to allow low energy X-rays and alpha particles to enter the active region of the detector. The cathode is formed by a 10 μm thick aluminized Kapton foil, that furthermore seals the holes in the detector lid. Gas and cathode high-voltage are fed into the detector through the detector lid, avoiding a thin feedthrough in the aluminum gas frame.

A stable 10 mm thick aluminum base plate ensures flatness of the readout structure. The detector components are screwed together, gas-tightness is ensured by 2.5 mm Viton O-rings.

Groups of ten anode strips are individually connected via exchangeable 0603 SMD resistors outside the active volume to a common high-voltage distribution point, that is again connected via a small 1 kΩ resistor to the global detector high-voltage. On the other side of the strips, the signal is decoupled using high-voltage resistant 0805 SMD capacitors. A 130 pin Panasonic connector[6] (Panasonic Corp. 2013) allows for connecting the readout electronics.

For cosmic muon tracking measurements under background irradiation with 20 MeV protons (Sect. 5.3) a second floating strip Micromegas has been constructed, see Fig. 3.15. The design parameters of the readout structure, carried by a 0.5 mm thick printed circuit board, are described above. Two $9 \times 64\,mm^2$ windows, closed by 10 μm thick aluminized Kapton foil, are milled into the 15 mm thick aluminum gas frame. Protons can completely cross the detector in a plane, parallel to the anode structure, on tracks, perpendicular to the readout strips.

A $6.4 \times 6.4\,cm^2$ copper plane on the inside of the FR4 detector lid forms the cathode plane. The gas is fed into and released from the detector through holes in the gas frame with 2.5 mm diameter. As above, the stainless steel mesh is glued to the aluminum gas frame. In order to reduce multiple scattering of traversing cosmic

---

[6]AXK6SA3677YG.

muons, the readout structure is supported by a light but stable base plate, consisting of aluminum honeycomb, sandwiched by two 0.5 mm thick FR4 layers.

3.4.2 6.4 × 6.4 cm² Floating Strip Micromegas Doublet with Low Material Budget

For tracking measurements with low energy ions, a semi-integrated floating strip Micromegas doublet with low material budget has been developed. It was used for 20 MeV proton tracking measurements at the tandem accelerator in Garching (Sect. 6.1) and for hydrogen and carbon ion tracking experiments at the Heidelberg Ion Therapy center (Sect. 6.2).

The internal setup of the detector doublet is shown in Fig. 3.16. Two Micromegas detectors are mounted back to back in a single unit. In the following, the components of one Micromegas unit are described.

The 6.4×6.4 cm² active area is formed by 128 copper anode strips with 500 µm pitch and 300 µm width, carried by a 125 µm thin FR4 sheet. Individual strips are connected to high-voltage via 22 MΩ SMD resistors outside the gas filled volume. Signals are decoupled from the anode strips over a second layer of copper strips on the bottom side of the thin readout PCB. These readout strips are routed to a 130 pin Panasonic connector, allowing for the connection of e.g. APV25 based readout electronics (Sect. 4.1.2). A single pair of anode and readout strips form an intrinsically high-voltage sustaining capacitor with a coupling capacitance of 6.7 pF. The photo-structurable etch resist on the thin copper-clad FR4 material, used to construct the readout structure, is of rather bad quality, showing cracks and inclusions. They cause stripline interruptions on the readout structure of the lower detector, that has been produced first. Due to this, a few groups of inefficient strips are observed in the

Fig. 3.16 Schematic setup of the floating strip Micromegas doublet with low material budget

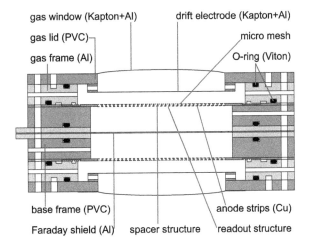

gas window (Kapton+Al)   drift electrode (Kapton+Al)

gas lid (PVC)   micro mesh

gas frame (Al)   O-ring (Viton)

base frame (PVC)   anode strips (Cu)

Faraday shield (Al)   spacer structure   readout structure

second layer. During the later production of the first layer, these defects have been patched by hand.

The approximately 150 µm thick spacer structure between anode strips and micromesh is photo-lithographically etched from solder resist film. Due to an already exhausted developer, the mesh-supporting pillars of the second layer are slightly swollen, resulting in an amplification gap width of (164 ± 1) µm. This problem has been avoided during construction of the first layer, that has been built afterwards, resulting in an amplification gap width of (150 ± 1) µm. The thin and flexible readout structure is glued onto a 10 mm thick PVC frame, with a $7.1 \times 7.1 \, \text{cm}^2$ cutout in the center. To ensure flatness, the gluing process was performed on a precise granite table.

The woven stainless steel mesh with 400 lines per inch and 25 µm thick wires is glued to the 10 mm thick aluminum gas frame. Gas is guided by two through-holes with 2.5 mm diameter into and out of the detector.

The gas volume is closed by a 10 µm thick aluminized Kapton foil, glued to the PVC gas lid. Due to the overpressure in the detector, this gas window is considerably bulged. The 6 mm wide drift region is formed by the micro-mesh and a drift electrode, glued to the gas lid. The drift electrode consists of a 10 µm thick aluminized Kapton foil and is contacted over a thin cable through the gas lid. Holes in the support structure of the drift electrode allow for a leveling of the gas pressure above and below the electrode, such that the cathode remains flat.

A (28 ± 2) µm thick aluminum foil shields the two Micromegas structures from each other, two small holes in the foil allow for a pressure equalization. In order to ensure a homogeneous amplification gap width, the space between the two readout structures is filled with detector gas at a relative overpressure of around 8 mbar with respect to the active regions of both Micromegas. The overpressure is dynamically created, by forcing the detector gas through a thin cannula. Thus the flexible readout structures are reliably pressed against the stretched and relatively rigid micro meshes. In measurements, no deformation of the drift gap width has been observed, slight periodic variations of the amplification gap width on the order of 2 µm are visible.

In order to avoid the extra capacitance and inductance of stripline vias, the readout connectors are mounted on the same side as the readout strips.

3.4.3 $48 \times 50 \, \text{cm}^2$ Integrated Floating Strip Micromegas

A large, integrated bulk floating strip Micromegas detector with an active area of $48 \times 50 \, \text{cm}^2$ has been constructed as a feasibility study. It has been tested with high energy pion beams produced by the Super Proton Synchrotron (SPS) at CERN, Sect. 5.2.

In Fig. 3.17 the detector is shown in plan view.

Its active area is formed by 1920 copper anode strips with 500 mm length, 250 µm pitch and 150 µm width. The strips are individually attached over printed resistors with approximately 10 MΩ resistivity to the high-voltage. This value is rather low, for

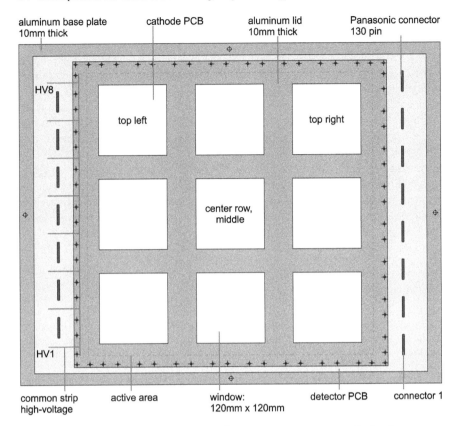

Fig. 3.17 Schematic view of the $48 \times 50\,\mathrm{cm}^2$ floating strip Micromegas. In the detector tests, discussed in Sect. 5.2, three different positions in the detector have been investigated: the top left, top right and the middle window of the center row

an optimized version, the resistivity will be at least doubled. The common strip high-voltage is subdivided into eight separate groups. Seven groups supply 256 strips each and one group supplies 128 strips with high-voltage. Signals from the anode strips are capacitively coupled onto a second layer of congruent readout strips. The strip layers are separated by a 75 µm thick FR4 layer. The readout structure is produced by first etching the anode strips including the footprint for the used 130 pin Panasonic connectors, and second, laminating the 75 µm copper-clad prepreg material onto the 2 mm thick printed circuit board, that carries the readout strips. The anode strips and the high-voltage distribution combs are etched in a second step.

The mesh and the corresponding support layers are laminated onto the readout structure, the support structure is then photo-lithographically produced in a single step. Thus the mesh is permanently attached to the readout structure. Detectors, fabricated with this method, are called bulk Micromegas (Giomataris et al. 2006).

The drift gap is defined by a 6 mm thick aluminum frame, sealed by 2.5 mm Viton O-rings. It is closed by a 1 mm thick copper-clad FR4 board, the cathode is formed by a $48 \times 50 \, cm^2$ copper plane on the inside of the detector lid. Cathode high-voltage is fed into the detector through a via in the FR4 detector lid. Gas can be flushed through the detector via four borings on two sides of the gas frame.

In order to reduce bulging, caused by gas overpressure in the detector, the Micromegas is sandwiched by two 10 mm thick aluminum plates, into which nine $120 \times 120 \, mm^2$ windows have been cut. It is shown in Sect. 5.2.4, that the drift gap shows nevertheless a variation on the order of 2 mm due to bulging. Especially in the middle of the cutouts, a considerable deviation from planarity is observed for a relative pressure of 30 mbar. It is foreseen to replace the aluminum covers by light but stiff FR4-aluminum-honeycomb sandwiches. In applications without space constraints, the cathode and the readout plane should be decoupled from the outer, gas-tight skin of the detector, similar to the design, discussed in Sect. 3.4.2.

References

Alexopoulos T, Burnens J, de Oliveira R, Glonti G, Pizzirusso O, Polychronakos V, Sekhniaidze G, Tsipolitis G, Wotschack J (2011) A spark-resistant bulk-micromegas chamber for high-rate applications. Nucl Instrum Meth A 640(1):110–118. doi:10.1016/j.nima.2011.03.025. http://www.sciencedirect.com/science/article/pii/S0168900211005869. ISSN 0168–9002

Bay A, Perroud JP, Ronga F, Derr J, Giomataris Y, Delbart A, Papadopoulos Y (2002) Study of sparking in micromegas chambers. Nucl Instrum Meth A 488(1–2):162–174. doi:10.1016/S0168-9002(02)00510-7. http://www.sciencedirect.com/science/article/B6TJM-45BRTKJ-V/2/a284420c8f18198d97bd3ecbd76666d4. ISSN 0168–9002

Bilevych Y, Blanco Carballo VM, Chefdeville M, Colas P, Delagnes E, Fransen M, van der Graaf H, Koppert WJC, Melai J, Salm C, Schmitz J, Timmermans J, Wyrsch N (2011) Spark protection layers for CMOS pixel anode chips in MPGDs. Nucl Instrum Meth A 629(1):66–73. doi:10.1016/j.nima.2010.11.116. http://www.sciencedirect.com/science/article/pii/S016890021002663X. ISSN 0168–9002

Bortfeldt J, Biebel O, Hertenberger R, Lösel Ph, Moll S, Zibell A (2013) Large-area floating strip micromegas. PoS, EPS-HEP2013:061

Byszewski M, Wotschack J (2012) Resistive-strips micromegas detectors with two-dimensional readout. JINST 7(02):C02060. http://stacks.iop.org/1748-0221/7/i=02/a=C02060

Campbell M, Chefdeville M, Colas P, Colijn AP, Fornaini A, Giomataris Y, van der Graaf H, Heijne EHM, Kluit P, Llopart X, Schmitz J, Timmermans J, Visschers JL (2005) Detection of single electrons by means of a micromegas-covered MediPix2 pixel CMOS readout circuit. Nucl Instrum Meth A 540(23):295–304. doi:10.1016/j.nima.2004.11.036. http://www.sciencedirect.com/science/article/pii/S0168900204024428. ISSN 0168–9002

Chefdeville M, Colas P, Giomataris Y, van der Graaf H, Heijne EHM, van der Putten S, Salm C, Schmitz J, Smits S, Timmermans J, Visschers JL (2006) An electron-multiplying micromegas grid made in silicon wafer post-processing technology. Nucl Instrum Meth A 556(2):490–494. doi:10.1016/j.nima.2005.11.065. http://www.sciencedirect.com/science/article/pii/S0168900205021418. ISSN 0168–9002

Chefdeville M, van der Graaf H, Hartjes F, Timmermans J, Visschers J, Blanco Carballo VM, Salm C, Schmitz J, Smits S, Colas P, Giomataris I (2008) Pulse height fluctuations of integrated micromegas detectors. In: Proceedings of the 9th international workshop on radiation imaging detectors. Nucl Instrum Meth A 591(1):147–150. 2007 doi:10.1016/j.nima.2008.03.045. http://www.sciencedirect.com/science/article/pii/S0168900208004221. ISSN 0168–9002

Galán J, Attié D, Ferrer-Ribas E, Giganon A, Giomataris I, Herlant S, Jeanneau F, Peyaud A, Schune Ph, Alexopoulos T, Byszewski M, Iakovidis G, Iengo P, Ntekas K, Leontsinis S, de Oliveira R, Tsipolitis Y, Wotschack J (2013) An ageing study of resistive micromegas for the HL-LHC environment. JINST 8(04):P04028. http://stacks.iop.org/1748-0221/8/i=04/a=P04028

Giomataris Y, De Oliveira R, Andriamonje S, Aune S, Charpak G, Colas P, Fanourakis G, Ferrer E, Giganon A, Rebourgeard Ph, Salin P (2006) Micromegas in a bulk. Nucl Instrum Meth A 560(2):405–408. doi:10.1016/j.nima.2005.12.222. http://www.sciencedirect.com/science/article/B6TJM-4J3NWY9-4/2/e3e28d951abf9e4f0afa59b7509c08f8. ISSN 0168–9002

Michael GmbH HJ (2013) TU-10M-8 Sicherheitsdatenblatt. unpublished

Jeanneau F, Alexopoulos T, Attie D, Boyer M, Derre J, Fanourakis G, Ferrer-Ribas E, Galan J, Gazis E, Geralis T, Giganon A, Giomataris I, Herlant S, Manjarres J, Ntomari E, Schune P, Titov M, Tsipolitis G, Wang W (2012) Performances and ageing study of resistive-anodes micromegas detectors for HL-LHC environment. IEEE Trans Nucl Sci 59(4):1711–1716, Aug 2012. doi:10.1109/TNS.2012.2198492. ISSN 0018–9499

Krieger C, Kaminski J, Desch K (2013) InGrid-based X-ray detector for low background searches. Nucl Instrum Meth A 729(0):905–909. doi:10.1016/j.nima.2013.08.075. http://www.sciencedirect.com/science/article/pii/S0168900213012163. ISSN 0168–9002

Linear Technology Corporation (2010) LTspice user's guide. http://www.linear.com/designtools/software/,

Nührmann D (2002) Das komplette Werkbuch Elektronik, Band 1 und 2. Franzis

Ochi A, Homma Y, Takemoto T, Yamane F, Kataoka Y, Masubuchi T, Kawanishi Y, Terao S (2013) Development of micromegas using sputtered resistive electrodes for ATLAS upgrade. presentation, given at the RD51 collaboration meeting, Zaragoza, Spain

Panasonic Corp (2013) P5K, P5KS Series, Narrow pitch connectors. http://pewa.panasonic.com/,

San Segundo Bello D, van Beuzekom M, Jansweijer P, Verkooijen H, Visschers J (2003) An interface board for the control and data acquisition of the Medipix2 chip. In: Proceedings of the 4th international workshop on radiation imaging detectors. Nucl Instrum Meth A 509(13):164–170. doi:10.1016/S0168-9002(03)01566-3. http://www.sciencedirect.com/science/article/pii/S0168900203015663. ISSN 0168–9002

Tektronix Inc (2013) Mixed signal oscilloscopes—MSO4000b, DPO4000b series datasheet. http://www.tek.com

Chapter 4
Methods

In the following, rather technical chapter, an overview over the applied electronic readout systems is given. The reconstruction mechanisms and algorithms, that have been developed and implemented in the course of this thesis are discussed. Details of the used hardware and the analysis algorithms can be found in the appendix.

A Gassiplex based and an APV25 based Micromegas readout system are discussed in Sect. 4.1. A reliable method for offline synchronization of separately acquired data streams is introduced. The detector gas mixture and pressure control system is described in Sect. 4.2.

Reconstruction and analysis methods are described in the following sections. The signal, cluster and hit position reconstruction in Micromegas is discussed in Sect. 4.3. In Sect. 4.5 different methods for the determination of the spatial resolution are introduced. The determination of the detection efficiency is described in Sect. 4.6. Alignment of Micromegas detectors in a tracking system is discussed in Sect. 4.7. In Sect. 4.8 a method for track inclination reconstruction with a single detector plane is presented. The method is often called µTPC reconstruction. The underlying ideas, Hough transform based data selection methods and the determination of the angular resolution are discussed. A detector simulation is presented, allowing for a quantitative explanation and correction of the systematic deviations of the method. The inversion of the method allows for determination of the electron drift velocity. The direct determination by measuring the maximum electron drift time is discussed.

4.1 Readout Electronics

Following the discussion in Chap. 2, the ionization charge, created by charged particles or photons in the drift region, is multiplied in avalanche processes in the amplification region between mesh and anode strips. The signal, created by the electron charge and the drifting positive ions, can then be detected as negative charge signal on the readout strips and as positive charge signal on the mesh. The latter is affected

© Springer International Publishing Switzerland 2015

J. Bortfeldt, *The Floating Strip Micromegas Detector*, Springer Theses,
DOI 10.1007/978-3-319-18893-5_4

by the capacitances within the detector, which strongly influence the pulse height of the signal, see Sect. 2.5.

During several calibration measurements, single channel charge sensitive preamplifiers from Canberra (Canberra Industries, Inc. 2007) and from the LMU electronics workshop with similar performance have been used to read groups of interconnected strips.

For measurements where only the pulse height was relevant, a high-rate capable Meilhaus ME-4610 ADC card (Meilhaus Electronic GmbH 2011) has been used to record the signal after amplification and shaping by an Ortec 452 spectroscopy amplifier. In timing and pulse height studies, the complete signal shape has been acquired with a setup based on a CAEN VME 12 bit 1 GHz flash analog-to-digital converter V1729 (CAEN S.p.A. 2010). This readout electronics has been described in detail in Bortfeldt (2010, Chap. 4).

In order to benefit from the good spatial and multi-hit resolution that Micromegas offer, the anode strips have to be read out individually. As small 9 cm × 10 cm Micromegas already have 360 strips, this can obviously only be achieved by employing highly integrated preamplifier electronics such as the Gassiplex or the APV25 chip.

4.1.1 Gassiplex Based Readout Electronics

The applied Gassiplex based readout electronics system has originally been developed for readout of the cathode plane of the HADES[1] Ring Imaging Čerenkov detector (Kastenmüller et al. 1999) and is kindly provided by the E12 group at the Technical University of Munich. The analog circuit on the boards was modified to cope with the negative charge signals encountered in Micromegas.

The 16 channel, multiplexing Gassiplex chip comprises a charge-sensitive preamplifier and a shaper for each channel (Beusch et al. 1994). The shaping time of 650 ns allows for a sufficiently long delay between particle passage and the generation of a trigger signal. Upon reception of a trigger, the actual shaper signal is stored by a track and hold circuit and passed via a multiplexer to the output of the chip. Thus the Gassiplex chips provides one charge value per strip and trigger, without any timing information. The trigger signal should reach the Gassiplex readout system about 500 ns after particle passage, since the readout system introduces an internal delay of 150 ns. Thus the coincident signal from triggering scintillators, which is available around 50 ns after particle passage has to be delayed by 450 ns. In order to measure the shaper signal (Fig. 4.1) and thus optimize the trigger delay, the additional trigger delay can be varied.

Despite the good amplification factor of 10 mV/fC, detectors equipped with the Gassiplex readout electronics are due to the relatively long shaper integration time insensitive to high-frequency noise if properly grounded.

---

[1] High Acceptance DiElectron Spectrometer.

Fig. 4.1 Gassiplex shaper output, measured with 120 GeV pions at the H6 beam line at SPS/CERN. Since the Gassiplex chip provides only one charge value per strip and trigger, the shaper signal has been measured by varying the delay between particle passage and trigger signal on the chips. The trigger delay on the horizontal axis is equivalent to time

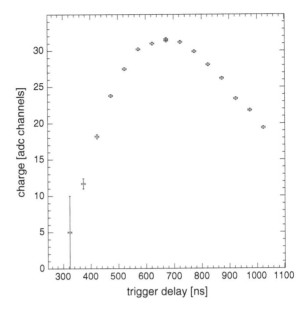

Further details about the preprocessing frontend modules and the VME based readout electronics can be found in the Appendix A.1.

4.1.2 APV25 Based Scalable Readout System

The 128 channel charge sensitive APV25 ASIC[2] has been developed as readout chip for the silicon micro-strip detectors in the tracker of the CMS experiment (Raymond et al. 2000). Each channel features a charge sensitive preamplifier, a shaping and a deconvolution circuit. The latter is intended for longer signals, specific to silicon detectors and is in general not used in Micromegas. A 192 cells deep pipeline per channel is used to continuously store analogue signals every 25 ns. Upon reception of a trigger signal, the chip consecutively outputs data from all 128 channels in a step function.

Three different modes of operation are possible: In peak mode, charge from a single column is output. In deconvolution mode, three consecutive cells per channel are read from the pipeline, weighted, added and multiplexed as a single value per channel to the output. In multi-mode, three consecutive columns can be read out. This feature can be used to measure complete pulse shapes by sending several triggers in succession. The APV25 operational parameters can be configured via a serial protocol.[3]

---

[2] Application-Specific Integrated Circuit.

[3] I^{2}C: Inter-Integrated Circuit communication.

As the analogue part as well as the output multiplexers run at the LHC bunch clock frequency of 40 MHz, the readout of e.g. 18 time bins for all 128 channels takes around 58 µs per chip. Although this is a non-negligible duration, the readout frequency of the APV25 based system is rather limited by the bandwidth of the 1 GBit Ethernet, over which data is passed to the data acquisition computer.

Responding to the need for a modular readout electronics the Scalable Readout System (SRS) (Martoiu et al. 2013) has been developed in the framework of the RD51 collaboration at CERN (Pinto 2010).

Further details about the Scalable Readout System and the APV25 frontend boards can be found in the Appendix A.2.

The trigger signal to the APV25 frontend boards is transmitted synchronously to the 40 MHz clock. As the triggering event is in usual applications not synchronized with the clock, the signal timing jitters within a 25 ns interval. For measurements in which the absolute timing is relevant e.g. reconstructing the cluster position with the µTPC method (Sect. 4.8), accuracy improves when correcting the signal timing for the 25 ns jitter. Details about the jitter correction are summarized in the Appendix A.2.

4.1.3 Merging of Data Acquired with Different Readout Systems

In a typical particle tracking setup, reference Micromegas detectors sandwich a Micromegas detector under test. The reference detectors are used to predict the particle track in the detector under test, in which predicted hit positions can be compared to internally measured hit positions to deduce spatial resolution and efficiency.

The additional temporal information of the APV25 system enables a wide variety of studies in the detector under test, such as investigation of single plane angular resolution or bunch spacing measurements. The reference detectors on the other hand have been designed for readout with the Gassiplex based system.

Since both systems have to be read out by separate DAQ computers, the two separate data streams have to be merged offline. Although both readout systems receive the trigger signal from the same source and runs can be started concurrently, a simple merging using the internal event counters is not feasible, since both systems, especially the Scalable Readout System, miss trigger blocks of variable length. A sufficiently exact event time stamp can be added to the Gassiplex event header in the readout software, since it is running on a real time Power PC. This is not possible in the Scalable Readout System due to the buffering of received UDP frames in the network interface controller in the DAQ computer.

In order to enable reliable offline merging, a so called Triggerbox has been developed, a 12 bit logic signal counter, that emits the current counter reading on 12 parallel logic signal outputs. These 12 logic signals are then duplicated and recorded in each readout system by a suitable module, see Fig. 4.2. In the VME based Gassiplex

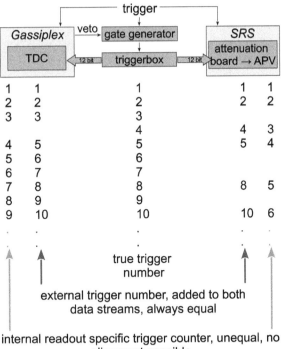

Fig. 4.2 Trigger signals are counted in a 12 bit counter called Triggerbox and emitted binary coded. They are recorded with a Time-to-Digital converter in the Gassiplex and an APV25 front-end board in the Scalable Readout System

system, a CAEN V775 16 channel Time-to-Digital converter is used, for the Scalable Readout System the logic signals are recorded, after appropriate attenuation, with a dedicated APV25 front-end board.

Since the Triggerbox output pulses have a fixed delay to the input trigger signal, they can also be used to correct for the 25 ns signal jitter, caused by the synchronous transmission of the trigger signal to the APV25 chips (Sect. 4.1.2).

4.2 Gas Mixture and Pressure Control System

In the measurements presented in this thesis the Micromegas detectors have been operated with $Ar:CO_2$ based gas mixtures. A custom built system, that is able to volumetrically mix up to four different gases, was used to mix the desired detector gases from pure Argon and Carbon Dioxide and to control the pressure and the gas flux in the detectors.

The pressure in the detector system could be regulated and held constant with an accuracy on the order of 1 mbar. Note that ambient temperature variations change the gas density, which has not been compensated by an adaptation of the gas pressure.

Usually an overall gas flow on the order of 2 ln/h and an absolute pressure of 1013 mbar is used.

A detailed description of the system and a schematic drawing can be found in the Appendix A.3.

4.3 Signal and Cluster Reconstruction

In this section the analysis of raw signals and the calculation of hit positions from raw signals is discussed.

Two multi-channel readout systems have been used in the measurements, presented in this thesis. A Gassiplex based system (Sect. 4.1.1) and the APV25 based Scalable Readout System (Sect. 4.1.2). Detector strips are read out with individual input channels, such that a particle position information can be deduced from the charge signal on neighboring strips.

Since the Gassiplex chip delivers a single charge value per strip and trigger, no temporal information can be recorded. A typical cosmic muon signal, acquired with the Gassiplex readout system can be seen in Fig. 4.3. For strip i only signals with a charge value $q_i > 3\sigma_{noise,i}$ are considered for cluster building, where $\sigma_{noise,i}$ is the standard deviation of the measured noise signals on strip i. A cluster is a group of neighboring hit strips, where usually one not-hit strip is allowed between two hit strips. It represents the overall signal of a traversing particle. The measured noise standard deviation for a Micromegas detector with 360 strips is shown in Fig. 4.4.

Fig. 4.3 Typical cosmic muon charge signal, acquired with the Gassiplex readout system in a 360 strip standard Micromegas. The muon creates a signal on five adjacent strips

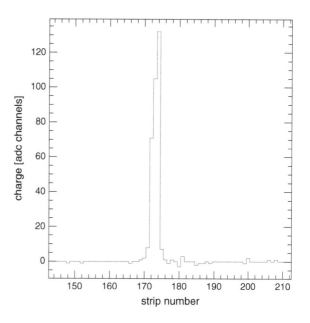

Fig. 4.4 Standard deviation of measured noise signal versus strip number in a 360 strip standard Micromegas during measurements with cosmic muons

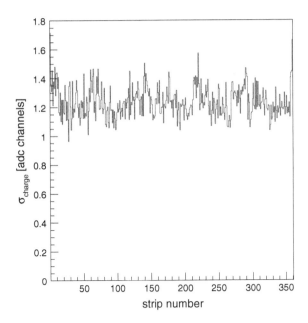

The total charge of the cluster q_{tot} is then given by the sum over all hit strips within the cluster:

$$q_{tot} = \sum_{i \in cluster} q_i \ .\tag{4.1}$$

The position in the detector-strip-coordinate system is then

$$x_{strips} = \frac{\sum_{i \in cluster} q_i \times i}{q_{tot}} \ ,\tag{4.2}$$

and can be translated into a space point

$$\vec{x} = \vec{p} + x_{strips} \, p_s \vec{d} \ ,\tag{4.3}$$

where $\vec{p}$ is the position, p_s the strip pitch and $\vec{d}$ the orientation of the detector in space.

Several particle hits i.e. clusters per event can be reconstructed in this way.

A typical APV25 raw signal on two neighboring strips can be seen in Fig. 4.5. By reading out 18 consecutive charge values with a temporal spacing of 25 ns, the signal evolution can be investigated. The total charge on strip i is given by the maximum height of the signal $q_{max,i}$, which is either defined by the charge value in the time bin with maximum charge or can be extracted by fitting the pulse with a skewed Gaussian

Fig. 4.5 Typical APV25
raw signal, measured with
protons in a 128 strip floating
strip Micromegas

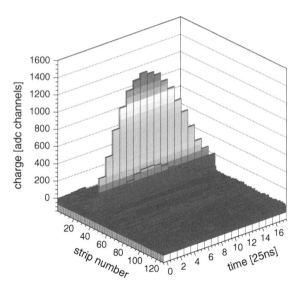

$$q(t) = q_{max} \exp\left(-\left(\frac{t^s - t_G}{t_w}\right)^2\right), \tag{4.4}$$

where $s \in]0, 1]$ defines the skewness of the Gaussian, t_w is a width parameter and t_G shifts the function in time. The position of the maximum is given by $t_{max} = t_G^{1/s}$.

Hit strips are as above identified by requiring $q_{max,i} > 3\sigma_{noise,i}$, the cluster charge is again given by the sum over all strip charges in the cluster. It has been seen though, that a better position reconstruction performance i.e. spatial resolution can be achieved by using the integrated charge signal

$$q_{int,i} = \sum_{t \text{ if } q_i(t) > \sigma_{noise,i}} q_i(t) \tag{4.5}$$

instead of q_i in the Eqs. (4.1)–(4.3).

In order to extract the signal timing more precisely than by using the position of the maximum t_{max}, the signal rising edge is fitted with an inverse Fermi function

$$q(t) = \frac{q_F}{1 + \exp\left((t_F - t)/t_{rise}\right)} + q_{baseline}, \tag{4.6}$$

where t_F is the point of inflection of the rising edge, t_{rise} is a parameter describing the signal rise time and $q_{baseline}$ represents the faintly fluctuating baseline before the signal, see Fig. 4.6. The signal height q_F and the rise time are constraint to realistic values $20 < q_F < 2380$ and $0.1 < t_{rise} < \infty$, the signal timing t_F is unconstrained in the fit. Since the time t_F is influenced by the signal rise time, which induces an additional time uncertainty, the rising edge is extrapolated onto the signal baseline:

Fig. 4.6 Signal fit with inverse Fermi function

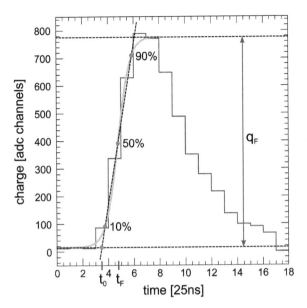

A straight line is constructed, that intersects with the inverse Fermi function at the points $0.1q_F$, $0.5q_F$ and $0.9q_F$. Its intersect with the baseline

$$t_0 = t_F - \frac{10 \ln 9}{9} t_{\text{rise}} \tag{4.7}$$

defines the signal timing (Lösel 2013).

Cluster positions are also calculated with the so called µTPC method (see Sect. 4.8 for details), which often yields better results for inclined tracks, where this charge-weighted-mean method fails.

Ionization by a perpendicularly incident particle and subsequent gas amplification creates a charge distribution in the amplification region with a typical width of 0.5–1 mm. If the pitch of anode strips is of the same order as the width of the Gaussian shaped charge distribution $q(x)$, discretization due to the periodic strip structure leads to a systematic mis-reconstruction of the particle hit position, Fig. 4.7.

The actual particle hit position is given by

$$x_{\text{true}} = \frac{\int xq(x)\mathrm{d}x}{\int q(x)\mathrm{d}x} . \tag{4.8}$$

Equation (4.8) is equivalent to Eq. (4.2) only in the limit of $n_i \to \infty$ i.e for a large number of strips in the cluster, or if x_{true} is exactly in the middle of a strip or exactly in the middle between two strips. In the vast majority of events $x_{\text{strips}} \neq x_{\text{true}}$.

For a hit cluster, consisting of two strips, the reconstructed hit position is shifted towards the middle between the two strips. For a three strip cluster, it is shifted

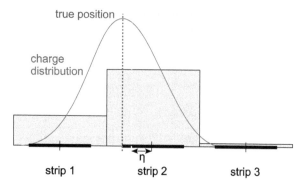

Fig. 4.7 Reconstruction of the hit position, defined by the Gaussian charge distribution (*red*), by measuring charge on three adjacent strips. The reconstructed hit position x_{strips} (*yellow*) is shifted towards the center of the three-strip cluster. The distance between the reconstructed position and the center of the nearest strip is defined as η

towards the middle of the central strip. This holds likewise for clusters formed by any even or odd number of strips, although this discretization mis-reconstruction quickly loses importance as soon as more than five strips are contributing to the cluster.

The distance between the reconstructed hit position and the center of the nearest strip is denoted by η and can be used to parametrize the mis-reconstruction. In Fig. 4.8 distributions of η are shown for two-strip and three-strip clusters. They were measured with a $6.4 \times 6.4 \, cm^2$ floating strip Micromegas doublet with 0.5 mm strip pitch in perpendicularly incident 88.83 MeV/u carbon ion beams.

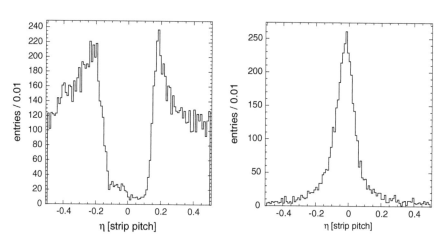

Fig. 4.8 Distribution of the difference between reconstructed hit position and the center of the nearest strip η for clusters consisting of two (*left*) and three (*right*) strips. Measured with perpendicularly incident 88.83 MeV/u carbon ions in a floating strip Micromegas with 0.5 mm strip pitch

For correct hit reconstruction, flat distributions are expected, due to the uncorrelated distribution of true hit positions within the irradiated area. The deviations from a flat distribution are due to the systematic mis-reconstruction. An η-dependent correction of the reconstructed hit position is possible that can be calculated iteratively from the η-distributions (Villa 2011). Different correction summands are calculated, depending on the number of strips n_i in the cluster.

For each bin k in the η histogram, the difference between the actual entry number and the expectation for a flat distribution $m_{\text{expect}} = n_{\text{entries}}/n_{\text{bins}}$ is determined

$$\Delta n_k = m_k - m_{\text{expect}} . \tag{4.9}$$

The correction $\Delta\eta_{n_i}(\eta)$ for a reconstructed hit $\eta = k/n_{\text{bins}} - 0.5$ from a cluster with n_i strips is then given by

$$\Delta\eta_{n_i}(\eta = k/n_{\text{bins}} - 0.5) = \frac{\sum_{j=1}^{k} \Delta n_k}{n_{\text{entries}}} . \tag{4.10}$$

In a second iteration, the measured hit positions are then corrected with the determined correction summands. In Fig. 4.9, the corrections are shown for hit clusters, that consist of two or three strips. They have been determined with the described method starting from the measured η distributions in Fig. 4.8.

This hit correction method has been used in measurements with floating strip Micromegas with 0.5 mm strip pitch (Sect. 3.4.2) in high-rate carbon ion and proton beams (Sect. 6.2), where an improvement of the spatial resolution on the order of 10 % has been achieved. In the other measurement campaigns that are discussed in this thesis, the improvements are considerably smaller, such that the hit correction has not been applied.

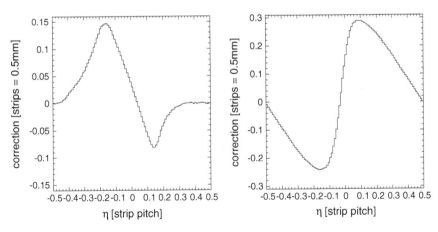

Fig. 4.9 Correction summands for reconstructed hit positions from clusters consisting of two (*left*) and three strips (*right*)

4.4 Track Reconstruction Algorithms

Two different algorithms have been implemented to identify and reconstruct straight tracks from measured particle hit positions in several Micromegas detectors. As background hits from photons or low energy particles can create additional hits in some or all detectors, the particle hit positions stemming from the true track must be found. For measurements where only one valid track per event is present, an iterative algorithm has been applied, that is called chain algorithm in the following. In high rate environments such as the tests at the Heidelberg Ion Therapy center (HIT, Sect. 6.2) with hit rates between 2 MHz and 2 GHz, considerably more than one particle track is detected per event. Under these circumstance the chain algorithm fails and a Hough transform based algorithm has been implemented.

Note that all Micromegas detectors included in track building, are required to be hit. This assumption can only be abandoned in tracking systems with more than three detectors and in events, in which a sufficiently large number of detectors ($\gtrsim 3$) registered the particle hit.

4.4.1 The Chain Algorithm

The chain algorithm is an iterative track building method, that can be used for reliable track reconstruction in conditions where only one valid track per event is present, but some or all detectors show background hits. The algorithm is sketched in the flow diagram Fig. 4.10.

Starting from the cluster with the highest charge in the first detector layer, a matching cluster in the second layer is searched, that lies within the geometric acceptance. If a cluster is found in the second layer, a straight line is fitted through the two cluster positions and the line is extrapolated into the next layer. Around the predicted position in this layer a matching cluster is searched. This is continued until reaching the final layer, where the clusters matching the track are now defined for all layers. If in any step no cluster is found in the expected region, the algorithm jumps back to the last layer and searches for alternative clusters in that layer.

4.4.2 Hough Transform Algorithm

The basics of the Hough transform algorithm are discussed in the Appendix B.1. A fast Hough transform based track finding algorithm has been developed, that is especially useful for track reconstruction in applications with high hit multiplicity.

All n_i hits (x_i, z_i) in all detector layers i are transformed into Hough space, using the Hesse normal form Eq. (B.3) as a transform function. The Hesse normal form describes a line in two dimensional z-x-space over their minimum distance to origin

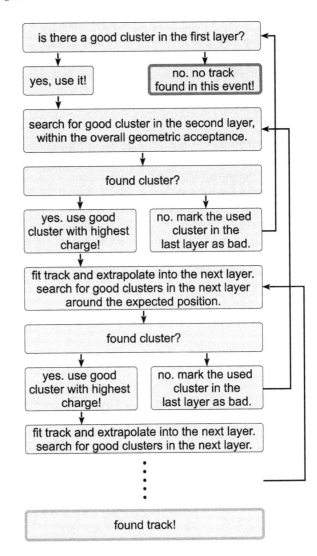

Fig. 4.10 Chain track finding algorithm

r and the angle α, enclosed by the line and the z-axis. The Hough space is spanned by r and α.

The detector layer with the minimum number of hits defines the maximum number of reconstructible tracks: $\min(n_i) = n_{\text{tracks,max}}$. The intersection points of all functions in Hough space are analytically calculated.

Intersection points are then cumulated in a physically correct way, until either the maximum number of tracks $n_{\text{tracks,max}}$ has been reconstructed or no valid intersection points are left anymore. The resulting cumulated points in Hough space define the track parameters of the reconstructed tracks. Physically correct means in this context, that the intersection points must lie inside a chosen track acceptance region, exactly one hit is used per detector layer, and a hit, that has been assigned to a specific track cannot be used a second time in an additional track.

The algorithm starts with the valid pair of intersection points, that have the smallest distance in Hough space, where the distance is defined by

$$\Delta(1, 2) = \sqrt{(r_1 - r_2)^2 + (c(\alpha_1 - \alpha_2))^2}. \tag{4.11}$$

The parameter c is used to scale the inclination difference $\Delta\alpha = \alpha_1 - \alpha_2$, such that it is correctly considered in the distance calculation. Otherwise the track distance term Δr dominates the distance calculation. For the setup used at HIT, a value of $c = 240$ mm has been chosen.

In Fig. 4.11 the Hough transform is shown for an exemplary event with three reconstructed ^{12}C ion tracks. In this event, three hits were reconstructed in the first detector layer, four in the second and four in the third. The track finding algorithm merged those hits into tracks, that were best compatible with straight tracks. All three tracks have been identified, the background hits in layer 2 and 3 do not spoil the track reconstruction.

In the measurements at the HIT (Sect. 6.2) up to seven tracks per event have been reconstructed.

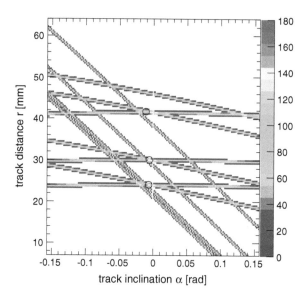

Fig. 4.11 Transformed hits from three detector layers, measured with 88.83 MeV/u ^{12}C ions with a beam intensity of 5 MHz at HIT. The *black circles* mark the parameters of the three reconstructed tracks. Functions with similar "inclination" correspond to coincident hits in the same detector layer. For didactic reasons, "thicker" functions are shown in the histogram, that allow for an identification of the correct track parameters (=intersection points) using the *color coding*. The absolute color scale is arbitrary. See Appendix B.1 for a more detailed discussion

4.4.3 Analytic Track Fitting

The chain and the Hough transform based track algorithms, discussed in Sects. 4.4.1 and 4.4.2, identify the correct combination of measured hits, that are best compatible with a straight track through the detector system.

The parameters of the track are then determined by fitting a line to the identified hit positions using a χ^2-minimization, Eq. (B.4). As discussed in the Appendix B.2, the track parameters for a linear function can be calculated analytically, without using iterative fitting algorithms. This results in a considerable acceleration of the track fitting.

Furthermore, the accuracy of the determined track can be calculated analytically from the known single detector spatial resolutions, Eq. (B.11). This is necessary in order to estimate the accuracy of a predicted hit position e.g. in the determination of the spatial resolution of a detector.

4.5 Determination of Spatial Resolution

The spatial resolution of a detector can be determined by comparing the true and the measured hit position in the detector. If this is done for many similar tracks, the measured hit position will scatter around the true hit position, the width of the distribution directly yields the spatial resolution. As the true hit position is often not known a priori, a set of reference detectors can be used to predict a hit position in the detector under test.

In the following, three different methods are discussed, that allow for determining the spatial resolution of a detector, based on particle tracks, measured in a system of at least three detectors.

4.5.1 Three Layer Method

The three layer method allows for determination of the spatial resolution in a detector system with at least four detectors. The detectors can have a different spatial resolution. It is an adaptation of the three wire method, that has been used in the OPAL jet chamber for the extraction of the spatial resolution (OPAL Collaboration 1984, 1991; Biebel et al. 1992).

Consider a tracking system consisting of four detectors, see Fig. 4.12. In a first step only the first three detector layers are considered.

A charged particle traverses the system, the particle hit positions in the first three detectors can be measured. The measured hit position r_1 and r_3 in layer MM1 and MM3 can be interpolated into the sandwiched layer MM2. For straight tracks, the expected hit position in this layer r_2' is then given by

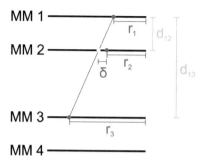

Fig. 4.12 Three layer method: a charged particle crosses (*blue line*) the detector system, consisting of four detectors (MM1–MM4). The particle hit positions (*blue circles*) in two layers (MM1 and MM3) can be used to predict the expected hit position (*yellow circle*) in a third layer (MM2)

$$r'_2 = (r_3 - r_1)\frac{d_{12}}{d_{13}} + r_1 , \qquad (4.12)$$

where d_{12} is the relative distance of MM1 and MM2 and d_{13} the distance between MM1 and MM3.

The residual between the predicted and the measured hit position in MM2 is

$$\delta = r_2 - r'_2 = r_2 - r_3\frac{d_{12}}{d_{13}} - r_1\left(1 - \frac{d_{12}}{d_{13}}\right) . \qquad (4.13)$$

Determining the residual δ for many tracks, yields a Gaussian shaped distribution of residuals, the width of this residual distribution is given by $\Delta\delta$. Denoting the spatial resolution of layer i by Δr_i and using Gaussian error propagation for Eq. (4.13), leads to

$$(\Delta\delta)^2 = (\Delta r_2)^2 + \left(\frac{d_{12}}{d_{13}}\Delta r_3\right)^2 + \left[\left(1 - \frac{d_{12}}{d_{13}}\right)\Delta r_1\right]^2 . \qquad (4.14)$$

This represents the relation between the spatial resolution of the three involved detectors and the measurable width of the residual distribution for tracks, defined by MM1 and MM3, interpolated into MM2.

For a system, consisting of four detectors, four different triplet combinations are possible,[4] such that we are left with four measurable widths $\Delta\delta_j$ and four unknown spatial resolutions $\Delta r_i = \sigma_{SR,i}$. The relation between the widths and the spatial resolutions is given by a system of four equations of the type shown in Eq. (4.14). Solving the system of equations, yields the four values of the spatial resolution.

For detector systems with more than four layers, the spatial resolutions can be determined by a χ^2 minimization of the over-determined equation system.

---

[4]MM1,MM2,MM3 and MM1,MM2,MM4 and MM1,MM3,MM4 and MM2,MM3,MM4.

4.5.2 Geometric Mean Method

An alternative method for the determination of the spatial resolution of position sensitive detectors has been proposed by Carnegie et al. (2005) and is presented in the following. A derivation can be found in the original publication.

Consider as above a system of four Micromegas detectors, that measure straight tracks of charged particles.

The measured hit positions in all four detectors can be fitted with a straight line in order to extract the track parameters. The residual between the measured and the predicted hit position in detector i can be determined for many similar tracks. The distribution of residuals can be fitted with a Gaussian function, yielding the width $\sigma_{in,i}$. The width $\sigma_{in,i}$ is smaller than the spatial resolution $\sigma_{SR,i}$ of detector i, since the detector is included in the track fit.

If detector i is explicitly excluded from the fit, the width $\sigma_{ex,i}$ of the resulting residual distribution is larger than the spatial resolution $\sigma_{SR,i}$, since the track is less constrained and the track prediction itself is afflicted with an error.

It has been shown by Carnegie et al. (2005), that the true spatial resolution of detector i can be approximated by the geometric mean

$$\sigma_{SR,i} \approx \sqrt{\sigma_{in,i}\sigma_{ex,i}} \ . \tag{4.15}$$

Using Eq. (4.15), the spatial resolution of all detectors in the system can then be approximated.

4.5.3 Track Interpolation Method

In a typical detector characterization measurement, particle tracks are measured by a set of reference detectors and extrapolated into a detector under test, see Fig. 4.13. The operational parameters of the detector under test are varied, while the reference detectors are operated under constant conditions.

Fig. 4.13 A track, measured in a set of reference detectors, is extrapolated into a detector under test. The residual Δx between the predicted and the measured hit position can be measured for many tracks and follows a Gaussian distribution

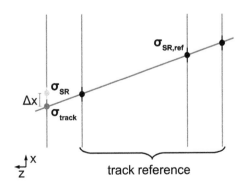

Assuming that the spatial resolution of the reference detectors has been previously determined with the three layer or the geometric mean method (Sects. 4.5.1 and 4.5.2), the accuracy of the hit position prediction σ_{track} in the detector under test can be calculated [Eq. (B.11) in the Appendix B.2].

Comparing the measured and the predicted hit position in the detector under test for many tracks, yields, as above, a distribution of residuals with standard deviation σ_{ex}. The spatial resolution of the detector under test is then given by

$$\sigma_{SR} = \sqrt{\sigma_{ex}^2 - \sigma_{track}^2} \ . \tag{4.16}$$

4.5.4 Comparison of the Three Methods

The spatial resolution of a $9 \times 10\,cm^2$ standard Micromegas for 120 GeV pions has been determined with the three different methods, described above. The drift field of the detector under test has been varied, the spatial resolution of the three reference detectors was $\sigma_{SR,ref} = (42 \pm 1)\,\mu m$.

The results are shown in Fig. 4.14 as a function of the drift field. The observed drift field dependence is discussed in detail in Sect. 5.1.4.

The spatial resolution, determined with the three layer and the track extrapolation method agrees within the respective error. Deviations by less than 2 μm are observed. It can safely be assumed, that the thus determined values are a good approximation of the true spatial resolution.

Fig. 4.14 Reconstructed spatial resolution for a $9 \times 10\,cm^2$ standard Micromegas, measured in 120 GeV pion beams with an Ar:CO$_2$ 85:15 vol% gas mixture. Superimposed is the mean spatial resolution of the reference detectors, operated at $E_{drift} = 0.3\,kV/cm$ (*blue line* and *error band*)

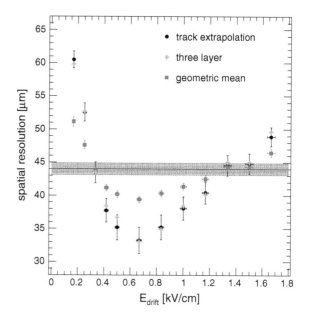

The values, extracted with the geometric mean method, deviate by as much as 9 µm. Agreement with the two other methods is only achieved at the drift fields, where the spatial resolution of the detector under test is close to the spatial resolution of the reference detectors.

The geometric mean method forces similar values for the spatial resolution for all detectors in the tracking system. It yields only reliable results, if the spatial resolution of the considered detectors is similar. It can e.g. be used to calibrate the spatial resolution of equal reference detectors, if they are operated at equal field parameters.

4.6 Detection Efficiency

The detection efficiency describes the ability of a detector to register the passage or hit of a particle or a photon. In tracking applications, a high efficiency to the traversing particles is desired whereas the efficiency to background hits should be as low as possible. Background radiation in high energy physics applications is usually composed of photons and neutrons.

In neutron tracking or photon detection applications, the efficiency to this radiation is enhanced by the use of a suitable solid or gaseous converter such as ^{10}B or high-Z materials like Gold or Xenon.

The efficiency can be calculated by comparing the number of registered hits in the detector n_{hit} to the number of particles or photons, that crossed the active area of the detector n_{tracks}:

$$\varepsilon = \frac{n_{hit}}{n_{tracks}} = \frac{n_{hit}}{n_{hit} + n_{no\,hit}} . \tag{4.17}$$

Since this is a counting experiment with two independent numbers n_{hit} and $n_{no\,hit}$, its error can then be calculated from Gaussian error propagation assuming Poisson statistic

$$\Delta\varepsilon = \sqrt{\left(\frac{\partial\varepsilon}{\partial n_{hit}}\Delta n_{hit}\right)^2 + \left(\frac{\partial\varepsilon}{\partial n_{no\,hit}}\Delta n_{no\,hit}\right)^2} = \sqrt{\frac{\varepsilon(1-\varepsilon)}{n_{hit} + n_{no\,hit}}} . \tag{4.18}$$

In tracking applications, the number of crossing particles can be determined by reference detectors, that sandwich the detector under test. The reference detectors can either be trigger scintillators or other Micromegas detectors. Using the latter often has the advantage, that the active areas are equal.

Assuming a Micromegas system, consisting of n detectors, the efficiency of the ith detector can be determined by counting the number of tracks registered in the other $n - 1$ detectors n_{ref} and comparing this to the number of tracks registered by all detectors n_{ref+i}:

$$\varepsilon_i = \frac{n_{\text{ref}+i}}{n_{\text{ref}}} = \frac{n_{\text{ref}+i}}{n_{\text{ref}+i} + n_{\text{all hit except }i}} \, . \tag{4.19}$$

This method yields reliable results. The determined efficiencies of the detectors are independent as long as inefficient spots in the reference detectors do not mask such spots in the detectors under test.

The so called hit efficiency is extracted without imposing any quality criteria on the track in the reference detectors by checking whether the detector under test registered *any* hit. If the reference detectors have a sufficiently good spatial resolution, the so called track efficiency can be determined by extrapolating the track prediction by the reference detectors into the detector under test and searching for a matching hit in the detector under test.

Note that the measured efficiencies, that are discussed later, are explicitly not corrected for dead time effect due to discharges or inefficient regions in the detector.

4.7 Detector Alignment

In a system consisting of several separate tracking detectors, the relative position of each detector has to be known to the µm level, if accurate track measurements are desired. Since the relative alignment can be externally measured with only sub-millimeter accuracy, tracks are used for the final accurate alignment.

Accurate alignment is relevant in application with more than two tracking detectors. In the procedure, we assume the position and the relative rotation of the outermost strip detectors to be known and constant and adjust the position and rotation of the inner detector layers. In the following the alignment with straight tracks in four consecutive steps is discussed, (1) the relative x-position, (2) the distance between the detectors, (3) the relative rotation around the z-axis and (4) the relative rotation around the y-axis. We only consider strip detectors with strips along the y-direction, that measure particle hit positions accurately in the perpendicular x-direction. For the alignment with respect to rotation around the beam axis (i.e. z-axis), additional hit position information in y-direction is necessary, which can either be provided by dedicated perpendicular detectors or by detectors with two-dimensional readout structure.

The difference between the hit position $x(z, y)$, predicted by the reference detectors, and the measured hit position x_{meas} in a certain layer is called residual and is given by

$$\Delta x = x(z, y) - x_{\text{meas}} \, . \tag{4.20}$$

The size and the sign of the thus defined residual is used in the track alignment procedure.

4.7.1 High-Precision Coordinate

Assume a system, consisting of three detection layers, where the middle layer is shifted in the direction of the precisely measurable hit position coordinate x, Fig. 4.15.

The difference between the intercept of the track, predicted by the outer layers, with the layer to be aligned and the reconstructed hit position in this layer (yellow circles in Fig. 4.15) is called residual. Measuring the residuals for many similar tracks, which are on average parallel to the z-axis, yields a distribution of residuals. The mean residual will be shifted with respect to the correctly aligned situation, where $\overline{\Delta x} = 0$. The residual shift $\overline{\Delta x} \neq 0$ directly gives the necessary position shift of the middle layer

$$\Delta \text{pos}_x = \overline{\Delta x} \,, \tag{4.21}$$

in order to correctly align the layer.

Note, that this method works of course also for the alignment in y-direction, if the detectors are position sensitive in y.

4.7.2 Distance

Track based alignment of the relative distance of detector layers is possible, if particles with at least two different inclinations are detected with the detector system. Assume again a system, consisting of three detector layers, where the middle layer is shifted in z-direction, see Fig. 4.16. First, consider tracks with positive inclination (blue line in Fig. 4.16). Due to the incorrect position of layer 2, negative residuals are observed $\overline{\Delta x} < 0$, see Eq. (4.20). Second, for tracks with negative inclination, positive residuals are measured $\overline{\Delta x} > 0$.

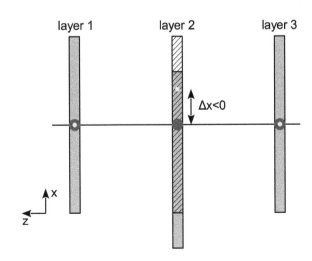

Fig. 4.15 Alignment method for the high-precision coordinate x. The true position of the middle detector is represented by the *filled rectangle*, the assumed is drawn *hatched*. True hit positions are marked with *red circles*, the reconstructed hit positions with *small yellow circles*

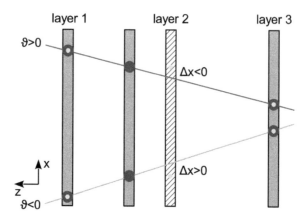

Fig. 4.16 The true position of the middle detector is represented by the *filled rectangle*, the assumed detector position is drawn *hatched*. Reconstructed particle hit positions are marked by *small yellow circles*, whereas the true hit positions are given by *red circles*. Tracks with positive inclination ϑ (*blue*) and negative inclination (*magenta*) lead to opposite sign residuals

The correlation between the residual and the track inclination angle ϑ allows for the direct determination of the necessary position correction: histogramming the residual as a function of $\tan(\vartheta)$ for different tracks, yields a linear correlation, that can be fitted with a straight line $\Delta x(\vartheta) = m \tan(\vartheta) + n$.

The intersect n is equivalent to the shift of the layer in x-direction and can be used to correct for it. The necessary position correction in z-direction is then given by

$$\Delta \text{pos}_z = -m \ . \tag{4.22}$$

4.7.3 Rotation Around the z-Axis

In the following, the alignment of layers with relative rotation around the z-axis is discussed. In order to be able to detect and then correct for relative rotation, the particle hit position must be measured in both dimensions, x and y. Imagine a three layer detector system, in which the middle layer is rotated by a small, positive angle $\phi > 0$ around the z-axis, Fig. 4.17. Note the definition of the sign of the rotation angle: positive rotation is defined as shown in the figure, a negative angle would correspond to a rotation towards the positive y-axis.

For tracks with positive y-hit positions, the reconstructed residuals are shifted towards positive values, $\overline{\Delta x} > 0$, and vice versa for tracks with negative y-hit positions. From a straight line fit of the histogrammed residuals as a function of the y-hit position with $\Delta x(y) = ky + j$, the rotation correction angle can be extracted:

$$\Delta \text{rot}_\phi = \arctan(k) \ . \tag{4.23}$$

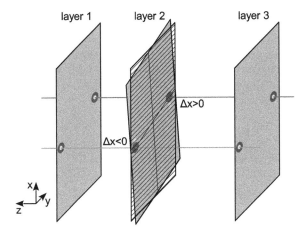

Fig. 4.17 Layer 2 is rotated with respect to the outer layers around the z-axis (*filled rectangle*). If this rotation is not accounted for (*hatched rectangle*), systematic deviations of the residuals as a function of the y-hit position are observed

4.7.4 Rotation Around the y-Axis

In measurements in which a detector is tilted with respect to the particle beam, the inclination angle of the detector has to be known. It can either be measured directly or can be deduced by comparing hit positions, predicted by reference detectors, with hit positions, measured in the inclined detector.

The underlying idea is demonstrated in Fig. 4.18.

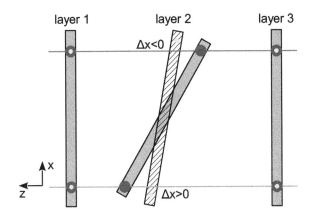

Fig. 4.18 Layer 2 is rotated around the y-axis, which is pointing into the paper plane (*filled rectangle*). This is the case e.g. in the measurements with tilted detectors, which allows for the investigation of the μTPC-reconstruction method. If the rotation angle is not correct (*hatched rectangle*), systematic deviations of the residuals as a function of the high accuracy x-hit position are observed

The middle detector is in reality rotated around the y-axis by a negative angle ϑ_y. Assuming a smaller inclination leads to a systematic linear dependence of the mean residual on the hit position in the high-precision coordinate x. For an assumed angle $\vartheta'_y > \vartheta_y$, the correlation between mean residual and x-hit positions is negative and vice versa. In the analyses, an iterative method has been used to correct for incorrect rotation around the y-axis.

4.8 Single Plane Angle Reconstruction—μTPC

In this section the reconstruction of the track inclination in a single Micromegas plane by measuring the arrival time of charge signals on strips is discussed.

4.8.1 The Method

Charged particles, traversing the 6 mm wide drift region of a Micromegas detector, create electron-ion-pairs along their way as discussed in Sect. 2.2. The electrons drift in the homogeneous drift field E_{drift} with a constant drift velocity $v_{\mathrm{d}} \sim 0.04$ mm/ns towards the amplification region and are detected there (Sects. 2.3 and 2.4). Those created close to the mesh, reach the amplification gap almost instantaneously, electrons created close to the cathode arrive approximately 150 ns later.

Since this method is inspired by the measuring principle of Time-Projection-Chambers e.g. in the ALICE experiment (ALICE Collaboration 2000), the method is also called μTPC reconstruction.

A typical event is displayed in Fig. 4.19. The arrival time t of charge signals, measured in bins of 25 ns, as a function of the corresponding strip number s is fitted with

$$t(s) = as + t_0 . \tag{4.24}$$

The slope a allows for direct calculation of the track inclination ϑ from the strip pitch p_s and the known drift velocity v_{d}

$$\vartheta = \tan^{-1}\left(\frac{p_s}{a \cdot v_{\mathrm{d}} \cdot 25 \text{ ns}}\right) . \tag{4.25}$$

The cluster position $s_0 = -t_0/a$ is given by the intercept of the line Eq. (4.24) with the horizontal axis and is for inclined tracks often more accurate than the position calculated from the charge-weighted mean (Sect. 4.3).

Fig. 4.19 Typical event, measured with 160 GeV pions with an inclination of 30°. The slope $\Delta t / \Delta$strip of the line fit (*red solid line*) allows for the direct determination of the track inclination ϑ

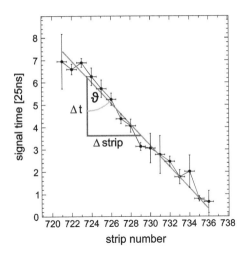

4.8.2 Data Point Selection and Weighting

The signal timing is extracted by fitting the strip signals within the selected cluster with an inverse Fermi function, as discussed in Sect. 4.3. The time-strip-data points are fitted with a straight line, its slope and intersect can be calculated analytically as discussed in Sect. B.2. By including information about the signal fit quality and plausibility in the line fit, the inclination reconstruction accuracy can be improved.

A signal timing distribution with the expected rectangular shape for tracks with 30° inclination is shown in Fig. 4.20. The reconstructed signal rise times are displayed in Fig. 4.21.

Initially the uncertainty $w_i = 1$ is assigned to all strip-time data points (see Eq. B.4). If the signal rise time parameter $t_{rise} < 0.35$ or $t_{rise} > 1.2$, the uncertainty is set to $w_i = 3$. Data points are completely ignored ($w_i = 10^4$) if the total charge on a strip, determined from a fit with a skewed Gaussian function Eq. (4.4), is too small i.e. $q_{max,i} < 3\sigma_{noise,i}$. The same applies if the pulse height from the Fermi signal fit Eq. (4.6) is below 10 ADC channels since then the fit has failed. Unphysical signals with rise times $t_{rise} < 0.2$ or $t_{rise} > 3$ and signal timing $t_0 < -2$ are also neglected.

If the residual from the line fit to weighted data points is not sufficiently small i.e. $\chi^2 > 2.5$, a Hough transform based algorithm (see Sect. B.1 in the Appendix) is used to identify data points not being compatible with a straight line.[5] These points are usually either mis-identified noise, signals distorted by δ electrons or signals from two particles, that were not separated in the cluster building process. Around each used data point in the time-strip-space a circle with radius 0.36 is drawn. 113 equidistant discrete points in each circle are transformed with the transform function

---

[5] Noise elimination using a Hough transform in the μTPC reconstruction has been proposed and developed in the framework of the Muon ATLAS Micromegas Activity collaboration (MAMMA) by G. Iakovidis, S. Leontsinis, K. Ntekas et al. The presented algorithm is inspired by this work.

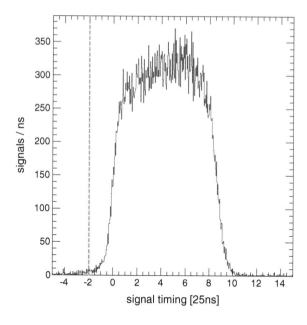

Fig. 4.20 Signal timing t_0 distribution for all hit strips, extracted from a Fermi fit to charge signals, measured with 120 GeV pions with an inclination of 30°. The signal timing is a measure for the electron drift time. Signals with $t_0 < -2[25\,\text{ns}]$ are excluded from the μTPC fit

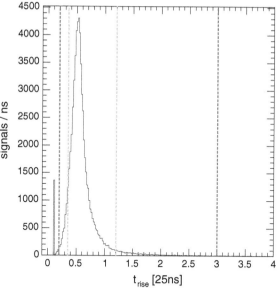

Fig. 4.21 Signal rise time parameter t_{rise}, measured with 120 GeV pions under an inclination of 30°. Data points from signals with a rise time between the *green lines* are considered with weight $w_i = 1$, signals between the *green* and *red lines* with weight $w_i = 3$ and signals outside the red lines are not considered in the μTPC fit

Eq. (B.3) into Hough space. In Figs. 4.22 and 4.23 a raw event and the respective Hough transform is shown.

By determining the bin with the maximum number of entries, the parameters r and α of the straight line, that is compatible with most points are found. The fit uncertainty of points with a distance larger than 1 mm to the line found are multiplied

Fig. 4.22 Signal time versus strip number, the final fit is displayed as *red solid line*. The accuracy of the point with $\Delta t = \pm 3$ has been downgraded due to an unplausible signal rise time, the two points with $\Delta t = \pm 10$ have been eliminated in the Hough transform signal cleaning

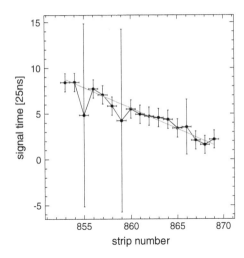

Fig. 4.23 Hough transformed data points. Clearly visible is the accumulation point at $(1.15, 8.0)$, where all but two curves intersect. The "thicker" functions allow for a simplified determination of the intersection point by searching the region with maximum number of entries (*red color*). The absolute color scale is arbitrary

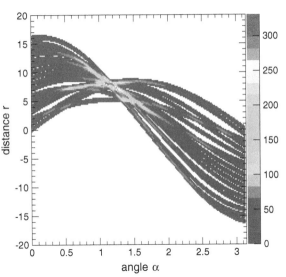

by 10. The points with now updated uncertainties are then fitted again with a straight line and the actual slope and intersect are determined.

4.8.3 Determination of Angular Resolution

The angular resolution and the reconstruction capabilities can be investigated in measurements with known track inclination e.g. by tilting the detector under test by a fixed angle with respect to the particle beam. A typical distribution of reconstructed

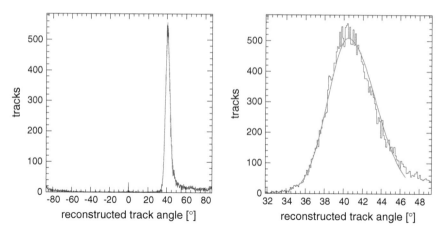

Fig. 4.24 Reconstructed track inclination for a true inclination of $(40\pm1)°$, measured with 20 MeV protons at $E_{amp} = 33.3$ kV/cm and $E_{drift} = 0.17$ kV/cm at the tandem accelerator Garching. The central peak is fitted with an asymmetric Gaussian Eq. (4.26) (*red* function). The entire angular range is shown on the *left* to demonstrate, that most reconstructed angles lie within the central peak. On the *right*, a zoom into the distribution shows the good agreement with the fit function

angles for a known track inclination of $(40 \pm 1)°$ can be seen in Fig. 4.24. The reconstruction capabilities are characterized by two parameters: The most probable reconstructed angle and the width of the distribution of angles.

The most probable reconstructed angle ϑ_{max}, given by the maximum of the angle distribution, is determined by fitting the upper part of the central peak with a Gaussian function. Misinterpretation of uncorrelated noise as signal and capacitive coupling of signals on neighboring strips lead to systematic deviations in the reconstruction, see Sect. 4.8.4. Thus the peak typically possesses tails to higher reconstructed angles. The resolution is determined by a fit with a piecewise Gaussian function

$$
\begin{aligned}
f(\vartheta) &= p_0 \exp\left(-0.5\left(\frac{\vartheta - \vartheta_{max}}{\sigma_<}\right)^2\right) \quad \text{for } \vartheta \leq \vartheta_{max} \\
f(\vartheta) &= p_0 \exp\left(-0.5\left(\frac{\vartheta - \vartheta_{max}}{\sigma_>}\right)^2\right) \quad \text{for } \vartheta > \vartheta_{max} ,
\end{aligned}
\tag{4.26}
$$

where p_0 is the height of the peak. The angular resolution is then given by the two asymmetric errors $\sigma_<$ and $\sigma_>$, representing the standard deviations of the piecewise Gaussian function.

Due to the non-linear transform function Eq. (4.25) between µTPC slope a and track inclination angle ϑ, histograms with variable bin width have to be used for a correct presentation of the track angle, see appendix B.3 for an example. Alternatively unbinned distributions can be analyzed.

4.8.4 Systematic Uncertainties of the Method

Additional to the finite angular resolution, caused by the finite accuracy in signal timing determination and non-homogeneous ionization, the reconstructed track inclinations are subject to systematic shifts. In the following section, the influence of the capacitive coupling of neighboring strips and the systematic mis-reconstruction of the cluster position at the borders of the charge clusters are discussed.

Both effects have been studied with a LTspice simulation, in which the Micromegas detector is modeled solely by the involved capacitances. The model is parameter-free, the applied capacity values have been calculated from the detector and strip geometry. In order to quantitatively compare simulated and measured values, the capacitance values of the $6.4 \times 6.4\,cm^2$ floating strip Micromegas doublet, examined in measurements with inclined 20 MeV proton beams (Sect. 6.1), have been used. The currents, produced by ionization and subsequent gas amplification, have been simplified as step functions, Fig. 4.25. Their length and timing is defined by the drift of ionization electrons in the drift region. The prolongation of the true signals, caused by the ion cloud, drifting from anode strips to the mesh, has been neglected. The total charge in each pulse is given by

$$Q = \int I(t)\mathrm{d}t = n_e G \,, \qquad (4.27)$$

where n_e is the number of ionization electrons and G the gas gain.

The simplified Micromegas model, used in the simulation, can be seen in Fig. 4.27. An explanation of the used symbol names is given in Table 4.1. Ideal pulsed current sources are assumed to inject charge onto the hit strips. Resulting charge signals are then detected via ideal charge sensitive preamplifiers.

Fig. 4.25 A traversing particle is assumed to homogeneously ionize the gas, producing a straight line of electrons (*dark blue line*), that drift into the amplification region. They are detected after gas amplification on five readout strips. The resulting currents on the strips are assumed to be constant during charge collection

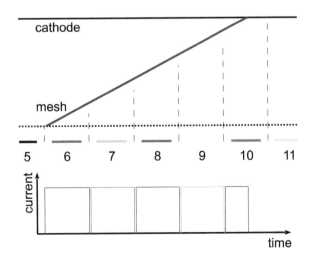

Table 4.1 Explanation and symbol names of the constituents of the LTspice detector model, Fig. 4.27

Name	Value	Description
C_{ms}	2.5 pF	Capacitance between mesh and anode strip, incorporating anode strip to ground capacitance
C_{asas}	4.1 pF	Capacitance between neighboring anode strips
C_{coupl}	7.0 pF	Coupling capacitance between an anode and the corresponding readout strip
R_{hv}	22 MΩ	Individual high-voltage recharge resistor
C_{slsl}	4.5 pF	Capacitance between neighboring strip lines, also including the readout strips
C_{slg}	1.9 pF	Capacitance between each strip line and ground
C_{cr}	1.0 nF	Coupling capacitance of readout electronics
C_{cc}	100 pF	Counter coupling capacitance of the ideal charge preamplifier

In Fig. 4.26 simulated charge signals for a true track inclination[6] of 20° can be seen. Due to the significant coupling between neighboring strips, even on the non-hit strip 5 a signal is detected with pulse height $ph_5 = 0.25\,ph_6$. The same is valid for strip 7, such that the signal on this strip begins to rise *before* ionization charge reaches the strip. This leads to a systematic shift of the signal timing for this and all following strips towards earlier values. A systematic overestimation of reconstructed track angles is the consequence, where the influence is largest for events with few hit strips, i.e. small track inclination.

Another systematic deviation comes from the mis-reconstruction of the charge position on the partially hit strip 10 (magenta strip in Fig. 4.25). The reconstructed charge position is shifted to the right with respect to the true position. Again, this leads to an overestimation of the reconstructed angle.

The resulting signals are fitted with an inverse Fermi function, their timing is determined by extrapolating the rising edge to the baseline, as discussed in Sect. 4.3. The resulting signal timings are compared with the true signal timings in Fig. 4.28. The offset between the reconstructed and the true signal timing for strip 6 is a systematic effect, caused by the description of the signal rise with a Fermi function. It would not have an effect on the reconstructed angle, if the deviations due to capacitive couplings would not be present. The shift of the reconstructed signal timing for strip 7, as described above, is clearly visible, for strip 8, the shift increases further and then tends towards a constant value.

---

[6]Note the definition of the track inclination as the angle between the vertical and the track, Fig. 4.19.

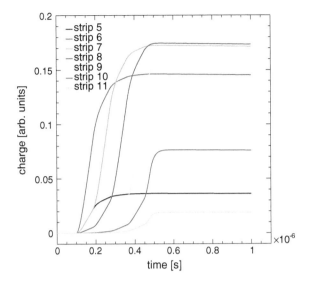

Fig. 4.26 Simulated charge signals on seven neighboring strips. The *color coding* is identical to Fig. 4.25

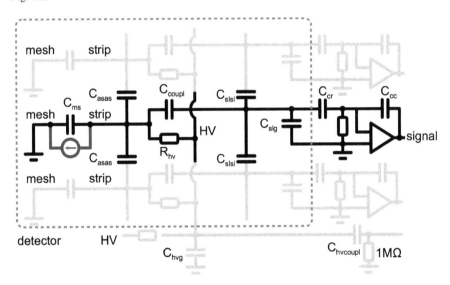

Fig. 4.27 Micromegas model, used in the LTspice simulation. The relevant capacitances and resistors for one strip/channel are shown, the neighboring strips are grayed out. An ideal current source (*blue*) is used to inject the signal onto the readout strip. The behavior of sixteen neighboring strips is fully simulated, correct boundary capacitances are used to also incorporate the influence of the other strips

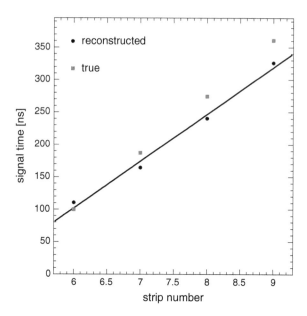

Fig. 4.28 Reconstructed signal times for a simulated track inclination of 20°, the *black line* represents the fit to the points. The true signal start times are superimposed. The systematic deviation of the signal timings is clearly visible

Four different track inclinations have been simulated, 10°, 20°, 30° and 40°. For each angle, four different equally spaced hit positions have been considered, to correctly include the edge effects. The resulting mean reconstructed track angles as a function of the true track inclination are shown in Fig. 4.29. For 10° track inclination a deviation of +9° is expected, the reconstructed angles approach the true values for increasing track inclination from above.

Simulated reconstructed angles are compared to measured angles in Sect. 6.1.5, showing a good agreement.

In applications, where the absolute reconstructed track inclination is *used* e.g. for measuring the scattering angles of beam particles in a scattering target, the reconstructed angles can be shifted with an angle-dependent scale factor to agree with the true track inclination. This calibration procedure is possible due to the monotony of the deviation towards larger angles as a function of the true track inclination. It can either be calibrated with data from inclined tracks, where the track inclination is known from e.g. a stack of tracking detectors, or it can be calibrated by fitting the predicted deviation, Fig. 4.29.

The latter approach will be shown exemplarily in the following. We search for the transform function $f(\vartheta_{\text{reco}})$, defined by

$$\vartheta_{\text{true}} = f(\vartheta_{\text{reco}})\,\vartheta_{\text{reco}}\,, \tag{4.28}$$

allowing for the calculation of the true track inclination ϑ_{true} from the reconstructed inclination value ϑ_{reco}. It can be found by fitting the inverse of Fig. 4.29, i.e. the true versus the measured track inclination with a suitable, empiric function

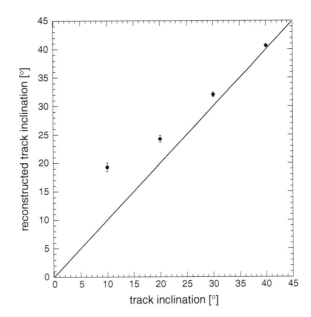

Fig. 4.29 Reconstructed track inclinations as a function of the true track inclination. The errors are calculated from the errors of the line fit to the strip-time-points. The *black line* represents the expectation without systematic deviations

$$\vartheta_{\text{true}}(\vartheta_{\text{reco}}) = \vartheta_{\text{reco}}\left(1 - \frac{p_0}{\vartheta_{\text{reco}}}\exp\left(-\frac{\vartheta_{\text{reco}}}{p_1}\right)\right), \qquad (4.29)$$

showing the correct asymptotic behavior for large angles.

Comparing Eq. (4.29) to (4.28) directly yields the correction factor $f(\vartheta_{\text{reco}})$, which has been plotted for the relevant range in Fig. 4.30.

Note, that the μTPC-reconstruction method does not work for angles below $10°$ in Micromegas with usual drift gaps on the order of several millimeters.

4.8.5 Inverting the Method—Determination of Electron Drift Velocities

In order to determine the track inclination angle from a fit with a straight line with slope a to the time-strip-data points using Eq. (4.25), the electron drift velocity v_d has to be known. It depends strongly on the detector gas composition and the electric drift field and weakly on gas temperature and pressure and can be calculated from microscopic scattering cross sections (Magboltz 2010) with the program MAGBOLTZ (Biagi 1999).

In the measurements with 120 GeV pions (Sect. 5.2) and with 20 MeV protons (Sect. 6.1) the detector is inclined relative to the beam particles such that the track inclination is known. By inverting Eq. (4.25) and using the known track inclination ϑ, the electron drift velocity can be determined

Fig. 4.30 Correction factor between the reconstructed and the true track inclination angle. The determined parameters, defined in Eq. (4.29) are $p_0 = (96 \pm 4)°$ for $p_1 = 8.112°$

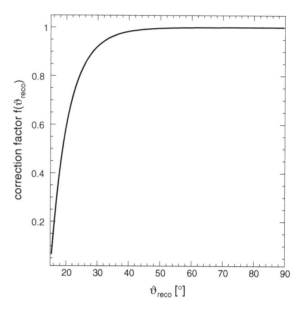

$$v_d = \frac{p_s}{a \tan(\vartheta) 25 \text{ ns}} . \qquad (4.30)$$

This method can be used to determine absolute drift velocities, but the systematic uncertainties of the µTPC reconstruction method have to be kept in mind: Due to capacitive coupling of adjacent strips, reconstructed angles for small track inclination are too large, for increasing track inclination, the difference decreases.

The systematic angle-dependent shift of the reconstructed angles towards larger angles is equally visible in the reconstructed drift velocity. Combining Eqs. (4.25) and (4.30) yields the relation between true and reconstructed angles and velocities

$$v_{d,\text{reco}} = \frac{v_{d,\text{true}}}{\tan(\vartheta_{\text{true}})} \tan(\vartheta_{\text{reco}}) . \qquad (4.31)$$

The reconstructed drift velocity $v_{d,\text{reco}}$ approaches for increasing track inclination the true drift velocity from above, in the same way as the reconstructed track slope $\tan(\vartheta_{\text{reco}})$ approaches the true particle track slope.

In an application, where the µTPC reconstruction method can be calibrated with tracks with known inclination, this drift velocity reconstruction can be used for calibration: The systematic deviations of the method can be absorbed in a (variable) effective drift velocity.

The drift gap width does not have to be known for reconstruction of the drift velocity with this method. When plotting the drift velocities versus the electric drift field, the drift gap width d_{drift} is the scale factor between applied drift voltage and computed electric field $E_{\text{drift}} = U_{\text{drift}}/d_{\text{drift}}$, see Fig. 4.31. By comparing the dependence and especially the position of the maximum of the computed drift velocities

Fig. 4.31 Electron drift
velocity for different drift
gap widths as a function of
the drift field for
$E_{amp} = 37.5\,kV/cm$.
Measured with 120 GeV
pions with an incidence
angle of $(29 \pm 1)°$ by
inverting the µTPC
reconstruction

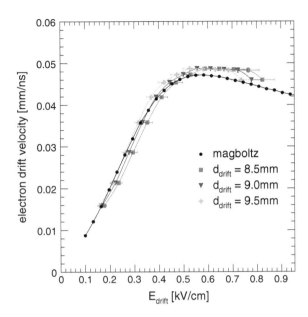

on the drift field with the values, determined with the method described, the drift gap width d_{drift} can be determined. This is especially useful to investigate distortions of Micromegas, operated with an elevated pressure with respect to the surrounding.

Note that systematic shifts of the reconstructed drift velocities occur, when a histogram with fixed bin width is used to represent and further analyze the reconstructed values. An explanation with example is given in the Appendix B.3.

4.8.6 Direct Determination of Electron Drift Velocities

The electron drift velocities can also be directly determined from analyzing the measured signal starting times. Since charged particles ionize the gas on average homogeneously along their path, all possible signal timings are equally probable, see Fig. 4.32. The maximum signal timing i.e. the maximum drift time t_{max} corresponds to ionization processes directly at the cathode. It can be extracted by fitting the edges of the distribution of signal timings with two Fermi functions. The beginning of the distribution is shifted to $t \sim 0$ by the jitter correction (see Sect. 4.1.2), its end is determined by extrapolating the falling edge with a straight line onto the baseline in the same way as used in signal fitting, see Sect. 4.3.

Since the electric drift field in the drift region between cathode and mesh with width d_{drift} is constant, the same holds for the drift velocity, which is then given by

$$v_d = \frac{d_{drift}}{t_{max}} . \tag{4.32}$$

Fig. 4.32 Signal timing distribution, fitted with two Fermi functions to extract the maximum drift time t_{max}

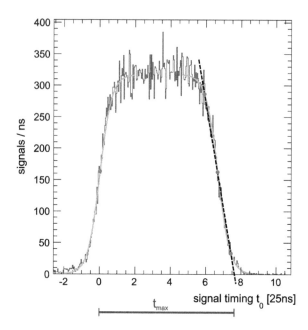

Thus the determined drift velocity is, in contrast to the method described in Sect. 4.8.5, not biased by the systematic uncertainties of the µTPC reconstruction method and provides an additional method to measure the drift velocity. By comparing the absolute values and the dependence on the drift field with the calculated theoretical drift velocities, the drift gap width d_{drift} can be determined, see Fig. 4.33. The limitation of the method becomes visible for low drift fields, where the signal timing distribution is cut due to the limited recording time window. Since the latest signals are simply missed, the maximum drift time appears to be too short.

For steep track inclinations or measurements with strongly ionizing particles such as low energy protons, where a large amount of charge is collected on only few strips and the ionization is dense, the maximum drift time is difficult to determine. Large charge clusters also create signals on neighboring strips such that the signal timing of charge clusters arriving later is distorted towards earlier values. The tail of the timing distribution is thus washed out.

It should be noted, that the two methods of drift velocity measurements, described in Sect. 4.8.5 and this section, are in principle independent. The main systematic uncertainty common to both, is the scale factor between the measured signal timings in "time bins" and in nanoseconds, which should be accurate on the per-mill level.

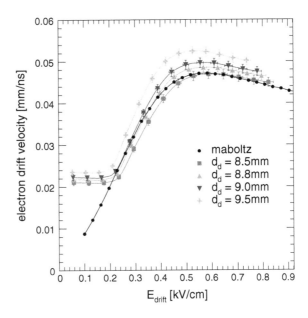

Fig. 4.33 Electron drift velocity for different drift gap widths as a function of the drift field, measured at $E_{amp} = 37.5\,kV/cm$. Determined from the maximum drift time t_{max} from 120 GeV pions with an incidence angle of $(29 \pm 1)°$. The three points deviating from the MAGBOLTZ simulation at low drift field are caused by a mis-reconstruction of the maximum drift time due to the limited recording time window

References

ALICE Collaboration (2000) ALICE time projection chamber: technical design report. Technical design report CERN-OPEN-2000-183, CERN, Geneva. http://cds.cern.ch/record/451098

Beusch W, Buytaert S, Enz CC, Heijne EHM, Jarron P, Krummenacher F, Piuz F, Santiard JC (1994) Gasplex a low noise analog signal processor for readout of gaseous detectors. Internal note, CERN-ECP/94-17

Biagi SF (1999) Monte Carlo simulation of electron drift and diffusion in counting gases under the influence of electric and magnetic fields. Nucl Instrum Methods A 421(1–2):234–240. ISSN 0168–9002. doi:10.1016/S0168-9002(98)01233-9. http://www.sciencedirect.com/science/article/pii/S0168900298012339

Biebel O, Boden B, Levegrün S, Rollnik A, Schreiber S, Kowalewski R, Behnke T, Breuker H, Hagemann J, Hansroul M, Hauschild M, Heuer R-D, Prebys E, Ros E, Van Kooten R, Binder U, Mohr W, Schaile O, Wahl C, Bock P, Bosch HM, Dieckmann A, Eckerlin G, Igo-Kemenes P, Tysarczyk-Niemeyer G, von der Schmitt H (1992) Performance of the OPAL jet chamber. Nucl Instrum Methods A 323(1–2):169–177. ISSN 0168–9002. doi:10.1016/0168-9002(92)90284-B. http://www.sciencedirect.com/science/article/pii/016890029290284B

Bortfeldt J (2010) Development of micro-pattern gaseous detectors—micromegas. Diploma thesis, Ludwig-Maximilians-Universität München. http://www.etp.physik.uni-muenchen.de/dokumente/thesis/dipl_bortfeldt.pdf

CAEN S.p.A. (2010) Technical Information Manual Mod. V1729. http://www.caen.it

Canberra Industries, Inc. (2007) Canberra 2004 data sheet. http://www.canberra.com

Carnegie RK, Dixit MS, Dubeau J, Karlen D, Martin J-P, Mes H, Sachs K (2005) Resolution studies of cosmic-ray tracks in a TPC with GEM readout. Nucl Instrum Methods A 538:372–383. ISSN 0168–9002

Kastenmüller A, Böhmer M, Friese J, Gernhäuser R, Homolka J, Kienle P, Körner H-J, Maier D, Münch M, Theurer C, Zeitelhack K (1999) Fast detector readout for the HADES-RICH. Nucl Instrum Methods A 433(1–2):438–443. ISSN 0168–9002. doi:10.1016/S0168-9002(99)00321-6. http://www.sciencedirect.com/science/article/B6TJM-3X64HB7-26/2/ab4be60e608f86382fea108ceb2ed8be

Philipp Lösel (2013) Performance studies of large size micromegas detectors. Master's thesis, Ludwig-Maximilians-Universität München. http://www.etp.physik.uni-muenchen.de/dokumente/thesis/master_ploesel.pdf

Magboltz (2010) Magboltz cross sections. http://rjd.web.cern.ch/rjd/cgi-bin/cross

Martoiu S, Muller H, Tarazona A, Toledo J (2013) Development of the scalable readout system for micro-pattern gas detectors and other applications. JINST 8(03):C03015. http://stacks.iop.org/1748-0221/8/i=03/a=C03015

Meilhaus Electronic GmbH (2011) Meilhaus Electronic Handbuch ME-4600 Serie 2.3D. http://www.meilhaus.de

OPAL Collaboration (1984) OPAL status report; 1984 ed. Technical report CERN-LEPC-84-17. LEPC-M-51, CERN, Geneva. http://cds.cern.ch/record/732033

OPAL Collaboration (1991) The OPAL detector at LEP. Nucl Instrum Methods A 305(2):275–319. ISSN 0168–9002. doi:10.1016/0168-9002(91)90547-4. http://www.sciencedirect.com/science/article/pii/0168900291905474

Pinto SD (2010) Micropattern gas detector technologies and applications, the work of the RD51 collaboration. In: Nuclear science symposium conference record (NSS/MIC). IEEE, pp 802–807. doi:10.1109/NSSMIC.2010.5873870

Raymond M, French M, Fulcher J, Hall G, Jones L, Kloukinas K, Lim L-K, Marseguerra G, Moreira P, Morrissey Q, Neviani A, Noah E (2000) The APV25 0.25 μm CMOS readout chip for the CMS tracker. In: Nuclear science symposium conference record, vol 2. IEEE, pp 9/113–9/118. doi:10.1109/NSSMIC.2000.949881

Villa M (2011) Resistive micromegas: a performance study. In: Presentation, given at the 8th RD51 Collaboration Meeting, Kobe, Japan, September 2011

Chapter 5
Floating Strip Micromegas Characterization Measurements

The aim of this thesis is to understand and describe the performance and properties of floating strip Micromegas, intended for high-resolution particle tracking at high-rates and in high-background environments.

In order to characterize the constructed floating strip detectors with respect to detection efficiency, spatial and angular resolution, a reliable and high-resolution reference track telescope is needed. A track telescope with an optimum track resolution below 20 μm has been developed, based on $9 \times 10 \, cm^2$ standard Micromegas. In order to optimize its performance, calibration measurements with a prototype system, consisting of four detectors, have been performed. The behavior in pion and muon beams has been studied, two different Argon based gas mixtures have been tested. The collected data allow for an implementation of signal and track reconstruction methods, analysis algorithms were developed and optimized. The calibration and optimization measurements of the telescope are presented in Sect. 5.1.

The track telescope enabled the detailed investigation of a $48 \times 50 \, cm^2$ floating strip Micromegas prototype in high-energy pion beams. The behavior and the inter-dependencies of pulse height, detection efficiency, spatial and single plane angular resolution and discharges have been studied. The homogeneity of the prototype was investigated. The floating strip Micromegas characterization measurements can be found in Sect. 5.2.

The performance of floating strip Micromegas in high-rate background environments has been studied, by laterally irradiating a $6.4 \times 6.4 \, cm^2$ muon sensitive floating strip detector with a 550 kHz, 20 MeV proton beam. The cosmic muon detection capabilities, the behavior of the spatial resolution and the stability with respect to discharges are presented in Sect. 5.3.

The prevention of ion induced space charge effects is one of the major advantages of Micromegas with respect to wire-based gas detectors. A small backdrift of positive ions from gas amplification processes into the drift region may only limit the high-rate capability of Micromegas structures, used for the readout of large drift volumes. The ion backdrift has been determined in a dedicated measurement with high-rate 20 MeV proton beams and can be found in Sect. 5.4.

© Springer International Publishing Switzerland 2015
J. Bortfeldt, *The Floating Strip Micromegas Detector*, Springer Theses,
DOI 10.1007/978-3-319-18893-5_5

The application of floating strip Micromegas detectors with low material budget in low-energy ion tracking and in therapeutic ion beam characterization measurements, can be found in Chap. 6.

5.1 Micromegas Based Track Telescope in High-Energy Pion and Muon Beams

In the following section calibration measurements with a track telescope in pion beams at the H6 and in muon beams at the H8 beam line of the Super Proton Synchrotron/CERN are described. The tracking system is intended for providing reference track information for detector tests in high-energy pion and muon beams. Excerpts of this section have been published by Bortfeldt et al. (2013b).

The track telescope has been tested with perpendicularly incident 120 GeV π^- at flux densities between 2 and 12 kHz/cm^2 at the H6 beam line. An Ar:CO$_2$ 85:15 vol% gas mixture has been used during these measurements. During the calibration measurements, the Gassiplex based Micromegas readout system has been optimized with respect to trigger delay and readout rate, particle induced discharge probabilities have been determined, efficiency and spatial resolution as a function of drift and amplification fields have been investigated. Since the pion beam spot size was only 3.5×1.5 cm^2 FWHM, the spatial resolution is independent of small relative rotations of the detectors.

The measurements with muons with 68.8 GeV$\lesssim E_\mu \lesssim 120$ GeV at the H8 beam line allowed for the determination of the detector homogeneity with respect to gas amplification and spatial resolution, due to the beam spot being larger than the detector area and thus yielding an approximately homogeneous illumination. Two different Ar:CO$_2$ mixtures, 85:15 and 93:7 vol%, were used. The detector performance as a function of drift and amplification field was studied. Several measurements were performed with a small inclination of the system with respect to the beam of $\pm 8°$ to study detector alignment. Due to the production of muons from decaying 120 GeV pions, the particle fluxes of approximately 3 Hz/cm^2 were three orders of magnitude lower than in the pion measurements. The muon energies are determined by the kinematics of the boosted two body decay $\pi^- \rightarrow \mu^- + \bar{\nu}_\mu$. A several meters long collimator in the beamline, that was used to stop the remaining pions, decreases the muon energy by less than 10 GeV.

The result discussion in the following sections follows the same scheme: Measurements with pions are presented first, in which the detectors were operated with an Ar:CO$_2$ 85:15 vol% gas mixture. The detector performance in muon and pion beams is then compared, for operation with the same gas mixture. Finally, measurements with muon beams and an alternative gas mixture of Ar:CO$_2$ 93:7 vol% are discussed.

5.1.1 Setup

The setup of the tracking telescope during calibration measurements with pions and muons is shown schematically in Fig. 5.1. Four standard Micromegas (Sect. 3.1.1) are mounted in a stable aluminum frame. The $9 \times 10\,cm^2$ active area of each Micromegas is subdivided by 360 copper readout strips with $250\,\mu m$ pitch and $150\,\mu m$ width. All readout strips ran along y-direction and thus measure particle hit position in x-direction. Coarse position information in the y-direction is given by two scintillator layers, consisting of three scintillators each. The readout system is triggered by coincident signals from both scintillator layers. The scintillators are read out with Hamamatsu R4124 photomultipliers (Hamamatsu Photonics K.K. 2010).

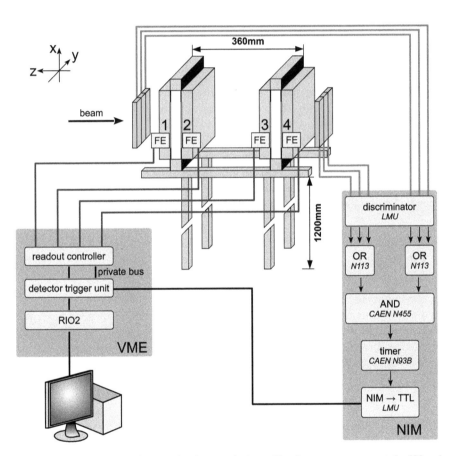

Fig. 5.1 Schematic setup of the track telescope during calibration measurements at the H6 and H8 beam lines at SPS/CERN. Four standard Micromegas detectors (*light blue*) with $9 \times 10\,cm^2$ active area (*red*) are mounted in an aluminum frame. 2×3 scintillators (*green*) provide the trigger signal for the readout system. The *right block* shows the trigger circuit, the *left* the schematic of the Gassiplex readout, which acquires charge signals from the Micromegas

Charge signals on the anode strips of the Micromegas detectors are acquired with the Gassiplex based readout electronics, described in Sect. 4.1.1. Six 64 channel front-end boards per detector are connected via a bus backplane with one port of the VME readout controller. Six of the 24 unused front-end board channels of the first detector are connected via an attenuator circuit to LEMO connectors and allow for the acquisition of logic signals with the Gassiplex readout system. This is used to acquire hit information from the six trigger scintillators.

Due to a faulty firmware on the front-end board FPGAs, the on-board threshold comparison was not working during the measurements. In order to reduce the data load on the Ethernet connection between the RIO2 VME controller and the DAQ computer, a software threshold comparison on the RIO2 was implemented, enabling a maximum trigger rate on the order of 650 Hz. In later measurements (Sects. 5.2 and 5.3), the on-board threshold comparison was fully operational.

Discharges between mesh and anode strips were registered by detecting the recharge signal at the mesh via a 470 pF capacitor. The recharge signals were amplified by Ortec 452 Spectroscopy Amplifiers and counted in a four channel scaler after discrimination and pulse elongation to 30 ms.

The detectors were constantly flushed with premixed Ar:CO_2 85:15 or with 93:7 vol% gas at a flux of 1.0 ln/h. The pressure was stabilized at (1010 ± 3) mbar absolute, using the gas system described in Sect. 4.2.

During the measurements with Ar:CO_2 85:15 vol% only three of the four Micromegas, namely 1, 2 and 4 were operable. Before switching to the 93:7 vol% mixture, Micromegas 3 was opened and repaired in a clean room. Due to a faulty cable connection, the perpendicular hit information from the scintillators was only recorded during the measurements with 93:7 vol%.

5.1.2 Pulse Height Behavior

In the following, the dependence of the pulse height of charge signals on the drift and amplification field is discussed. The term pulse height denotes the amount of *measured* charge on a cluster of adjacent hit strips (Sect. 4.3). It is related to the ionization charge q_0, the mesh electron transparency $t(E_{\text{drift}})$ (Sect. 2.6) and the gas gain $G(E_{\text{amp}})$ (Sect. 2.4)

$$ph(E_{\text{amp}}, E_{\text{drift}}) = cq_0 t(E_{\text{drift}}) G(E_{\text{amp}}) . \tag{5.1}$$

The proportionality factor $c = c_{q \to U} c_{\text{cap}}$ incorporates the charge-to-voltage conversion factor of the applied preamplifier circuit and furthermore a detector capacitance dependent factor, that has been discussed in detail in Bortfeldt (2010, Chap. 8). The pulse height of muon and pion signals is compared, the influence of the gas mixture on the relative gas gain and thus the pulse height is shown.

The measured pulse height distributions are fitted with a Landau function, convoluted with a Gaussian. The most probable value of the Landau function is then used

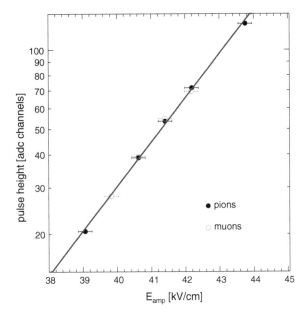

Fig. 5.2 Pulse height as a function of the amplification field in the first detector in pion and muon beams. Measured at $E_{drift} = 0.83\,kV/cm$ with an Ar:CO$_2$ 85:15 vol% gas mixture

for further pulse height analysis. The accuracy of the fit is used as vertical error in all figures in the following section. If no vertical error bars are visible, the errors are smaller than the markers. Systematic effects such as ambient temperature variation, have deliberately not been considered in the error calculation, in order to demonstrate their significance.

The pulse height as a function of the amplification field, measured with pions and muons in an Ar:CO$_2$ 85:15 vol% gas mixture, can be seen in Fig. 5.2. The expected exponential behavior, as discussed in Sect. 2.4, is clearly visible. Since the most probable energy loss and thus the ionization charge created by 120 GeV pions with $\beta\gamma = 860$ and ~90 GeV muons with $\beta\gamma = 850$ is equal (see Fig. 2.1), equal pulse heights are expected and also observed.

The pulse height as a function of the drift field for pions and muons, measured with Ar:CO$_2$ 85:15 vol%, is shown in Fig. 5.3. Typical Micromegas behavior is observed: For increasing drift field, the pulse height rises as the separation of electrons and positive ions, produced in ionization processes, increases. The initial pulse height increase is furthermore caused by the increasing electron drift velocity, and thus the decreasing signal rise time. If the signal rise time is larger than the fixed integration and shaping time of the preamplifier electronics, only a fraction of the produced charge is detected.

Maximum pulse height is reached for $E_{drift} \sim 0.7\,kV/cm$. The electronic transparency of the mesh and thus the pulse height decreases with further increasing drift field as more and more field lines end on the mesh and the transverse electron diffusion increases, which leads to a loss of ionization electrons on the mesh.

Fig. 5.3 Pulse height as a function of the drift field in the first detector for different amplification fields. Measured with pions and muons using an $Ar:CO_2$ 85:15 vol% gas mixture

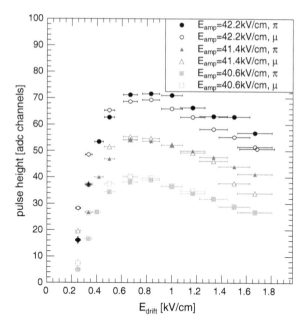

Again, equivalent behavior for pions and muons is observed, despite some small deviations that are due to ambient temperature variations between the measurements at the H6 beam line with pions and the H8 beam line with muons, two weeks later (discussion of the temperature dependence of gas gain in Sect. 2.4). The acquisition time for a single voltage scan was in the muon beam measurements considerably larger than in the pion measurements, due to the lower muon rate and the lower beam availability in muon measurements. Thus ambient temperature variation visibly influence the muon pulse height during single voltage scans.

The pulse height as a function of the amplification and the drift field, measured with muons, using an $Ar:CO_2$ 93:7 vol% gas mixture, is shown in Figs. 5.4 and 5.5 respectively. The absolute pulse height at equal amplification and drift fields is in 93:7 vol% by a factor of 3.0 $\pm$ 0.3 larger than in 85:15 vol%. Since the most probable ionization charge in 6 mm $Ar:CO_2$ 93:7 and 85:15 vol% is with $q_0 = 49.1\,e$ and $q_0 = 49.0\,e$ equal, the observed difference is caused by the higher gas gain in 93:7 vol%. This is consistent with the discussion in Sect. 2.4, where, according to the first Townsend coefficient, a gain difference by a factor of 2.8 is expected.

Additional to the different absolute scale, the maximum pulse height for 93:7 vol% is reached at a drift field of $E_{drift} \sim 0.4\,kV/cm$ as compared to $E_{drift} \sim 0.7\,kV/cm$ for 85:15 vol%, see Figs. 5.3 and 5.5. This can be understood by comparing the transverse electron diffusion coefficients for the two gas mixtures (cf. Fig. 2.5). An increased transverse electron diffusion leads to an increased loss of electrons at the mesh. The pulse height reaches a maximum, when the transverse diffusion is close to its minimum at $E_{drift} \sim 0.2\,kV/cm$ for 93:7 vol% and $E_{drift} \sim 0.4\,kV/cm$ for

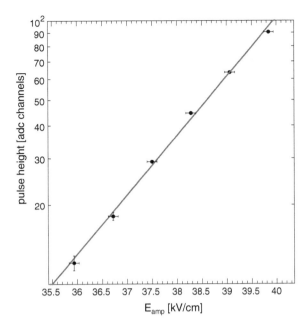

Fig. 5.4 Pulse height of muon signals as a function of the amplification field in the first detector. Measured at $E_{drift} = 0.83$ kV/cm with an Ar:CO_2 93:7 vol% gas mixture

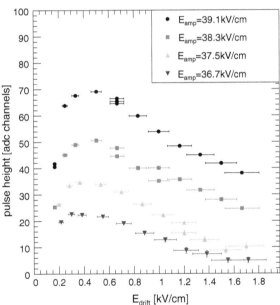

Fig. 5.5 Pulse height as a function of the drift field in the first detector for different amplification fields. Measured with muons using an Ar:CO_2 93:7 vol% gas mixture

85:15 vol%. The stronger decrease of the pulse height for increasing drift field, that is observed with the 93:7 vol% gas mixture, can in the same sense be correlated to a globally higher diffusion (cf. Fig. 2.5).

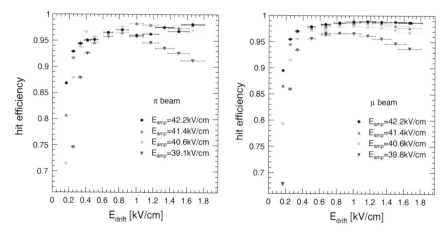

Fig. 5.6 Hit efficiency as a function of the drift field in the first detector for different amplification fields. Measured with pions (*left*) and muons (*right*) using an Ar:CO$_2$ 85:15 vol% gas mixture

5.1.3 Efficiency

The detection efficiency for pions and muons is compared in the following section. Its dependence on the drift and amplification field is discussed, the influence of the gas mixture is shown.

The detection efficiency for 120 GeV pions and approximately 90 GeV muons has been measured as a function of the drift and amplification field. It is determined with a tag-and-probe method, described in Sect. 4.6: The hit efficiency in the nth detector is determined by selecting events, in which the other $n-1$ detectors are hit. The events in this subset of events are then counted, in which also the nth detector registered a hit. The accuracy of the determined efficiency values is calculated, assuming Poisson statistics cf. Eq. (4.18). The markers in the figures in this section are larger than the errors, if no error bars are visible. The shown efficiency has intentionally not been corrected for systematic effects due to discharges and inefficient regions in the detectors.

The hit efficiency as a function of the drift field for pions and muons in an Ar:CO$_2$ 85:15 vol% gas mixture for various amplification fields is shown in Fig. 5.6.

Starting at low fields, the efficiency increases significantly with increasing drift field. This is closely connected to the improving separation of ionization charge and a decrease of the transverse electron diffusion in this region, as described in Sect. 5.1.2. The efficiency is furthermore influenced by the signal rise time, which is given by the sum of electron drift time in the drift region and ion drift time in the amplification region: If the signal rise time is longer than the preamplifier electronics integration time, only part of the produced charge is detected, the clustering of ionization leads to a considerable fluctuation of the detected charge and thus of the efficiency. The increase of the electron drift velocity with increasing field, leads to a decreasing signal rise time and thus a reduction of pulse height fluctuations, which are responsible for the low efficiency at low fields.

In pion beams the efficiency rises to values above 0.95 for $E_{drift} \gtrsim 0.5\,kV/cm$, for muons a plateau with values above 0.98 is reached for $E_{drift} \gtrsim 0.6\,kV/cm$. The maximum efficiency for perpendicularly incident minimum ionizing particles is limited to approximately 0.99 by the mesh supporting pillars, that cover 1.1 % of the active area. For further increasing drift fields the efficiency starts to decrease again, the effect is strongest for the smallest amplification fields. This is caused by the increasing electron mesh opacity and increasing transverse diffusion, that lead to a loss of ionization electrons at the mesh.

The efficiency is weakly correlated to the *absolute* pulse height, as can be seen nicely from the muon measurements. Optimizing the fraction of ionization charge, that reaches the amplification region and reducing pulse height straggling, has a strong influence on the efficiency.

An apparent difference between the efficiencies measured with pions and muons in $Ar:CO_2$ 85:15 vol% are the discontinuities and jumps, observed in the pion measurements. Since the measurement time for a single voltage scan for pions is, due to the higher rate and better beam availability, much shorter than for muons, this cannot be correlated with temperature fluctuations during the voltage scans. It is rather due to the higher discharge probabilities during the pion measurements and their impact on the detection efficiency.

In Fig. 5.7 the fraction of inefficient events is shown as a function of the event number during two measurements with pions and muons at equal electric field parameters. Clearly visible are the clusters of inefficient events during the pion measurement, that are caused by the inefficiency of the detector during the high-voltage recharge, following a discharge. During the shown muon measurement, no discharge was observed. The elevated baseline inefficiency for muon beam measurements is caused by the scintillator trigger acceptance, that is slightly larger than the Micromegas active area. This effect only appears in the muon measurements, where the trigger scintillators were almost homogeneously illuminated.

Discharge rates in the pion measurements are larger due to the globally higher particle fluxes and the composite structure of pions, that can create hadronic showers in the detector aluminum lid or the readout structure, leading to several highly-ionizing particles in the active volume. It should be noted, that the discharges are obviously non-destructive and that their influence on the efficiency is on the order of a few percent. Measured discharge probabilities are discussed in Sect. 5.1.5.

The detection efficiency for muons, using an $Ar:CO_2$ 93:7 vol% gas mixture, is displayed in Fig. 5.8. Its initial rise for small and later decrease for high drift fields is closely correlated to the amount of ionization charge, reaching the amplification region and detected during the preamplifier integration time, as discussed above. Efficiency plateaus above 0.985 are reached over a wide range of drift fields. The efficiency is in this case only limited by the mesh supporting pillars with $300\,\mu m$ diameter, that cover 1.1 % of the active area. Minimum ionizing particles, that hit pillars centrally are not detected.

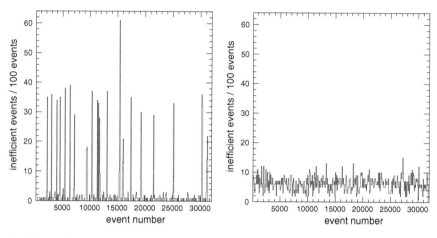

Fig. 5.7 Inefficient events per 100 events versus the event number in the first detector. Measured with pions (*left*) and muons (*right*) at $E_{amp} = 42.2$ kV/cm and $E_{drift} = 0.83$ kV/cm using an Ar:CO$_2$ 85:15 vol% gas mixture. Clearly visible are the clusters of inefficient events following a discharge in the detector in the pion beam and the absence of these clusters in the muon beam. An accepted scintillator trigger signals is counted as event in the two figures, regardless of whether the triggering particle actually crossed the active region of the Micromegas detectors. The slightly larger acceptance of the scintillator triggers with respect to the Micromegas active area gives rise to the elevated baseline inefficiency in the muon measurements

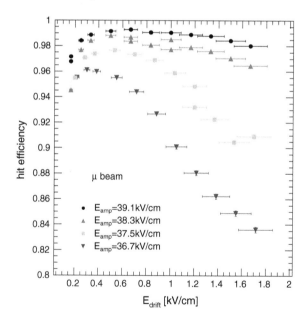

Fig. 5.8 Hit efficiency as a function of the drift field in the first detector for different amplification fields. Measured with muons using an Ar:CO$_2$ 93:7 vol% gas mixture. Note the different scale with respect to Fig. 5.6. Due to the higher gas gain in 93:7 vol% as compared to 85:15 vol%, lower amplification fields can be used

5.1.4 Spatial Resolution

The spatial resolution of the first Micromegas detector in the tracking system has been investigated as a function of drift and amplification field with two different $Ar:CO_2$ gas mixtures of 85:15 and 93:7 vol%, using 120 GeV pions and ~90 GeV muons. In Sect. 4.5 the methods, used to determine the spatial and the track resolution, are introduced.

The particle track, measured with the other reference Micromegas, is extrapolated into the first detector. Comparison of the predicted particle hit position with the hit position, measured in this detector, yields a distribution of residuals of width σ_{ex}. The spatial resolution of the first Micromegas can then be calculated from this width and the track accuracy using Eq. (4.16).

The track accuracy has been calculated from the spatial resolution $\sigma_{ref,i}$ of the reference detectors with Eq. (B.11). The reference detector spatial resolution has been determined with the geometric mean method (Sect. 4.5.2) in measurements, in which all detectors were operated with equal drift and amplification fields.

In Fig. 5.9 the spatial resolution of the first Micromegas is shown for measurements with pions and muons. The detector system was operated with an $Ar:CO_2$ 85:15 vol% gas mixture. The spatial resolution reaches a distinct minimum upon maximizing the fraction of ionization charge, that reaches the amplification region and is detected. It has been discussed in Sect. 5.1.2, that this is correlated to minimum transverse diffusion of electrons. Optimum values are observed at $E_{drift} = 0.65$ kV/cm with $\sigma_{SR} = (33 \pm 2)$ μm for pions and $\sigma_{SR} = (51 \pm 2)$ μm for muons.

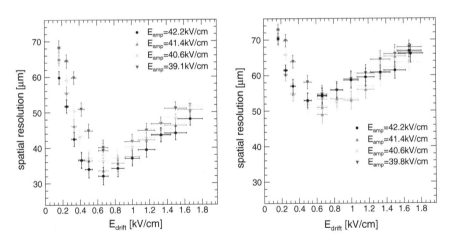

Fig. 5.9 Spatial resolution of the first detector as a function of the drift field for different amplification fields. Measured with pions (*left*) and muons (*right*) using an $Ar:CO_2$ 85:15 vol% gas mixture. The shown uncertainties of the spatial resolution are calculated from the uncertainty of the fit to the residual distribution and the uncertainty of the track accuracy

Fig. 5.10 Spatial resolution of the first detector as a function of the drift field for different amplification fields. Measured with muons using an Ar:CO$_2$ 93:7 vol% gas mixture. The y-hit information from the triggering scintillators is used to correct for relative detector rotation

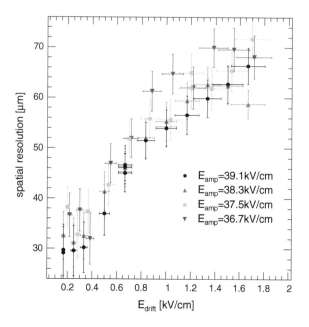

Although the shape of the spatial resolution as a function of the drift field is similar for pions and muons, the absolute values determined for muons are by $\approx$20 µm larger. This is due to the larger beam spot size in the muon measurements, where the whole active area of $9 \times 10\,\text{cm}^2$ was irradiated, in contrast to a beam spot area of only $3.5 \times 1.5\,\text{cm}^2$ FWHM during the pion measurements. Due to the larger beam spot, a relative rotation of the detectors degrades the observable spatial resolution. If, additional to the x-coordinate, that is measured precisely by the Micromegas, the y-coordinate of the particle hit position can be determined, a relative detector rotation can be corrected for on a track-by-track basis. During the muon measurements with the Ar:CO$_2$ 85:15 vol% gas mixture, the y-hit position was not acquired. It will be demonstrated below for the measurements with 93:7 vol%, where a coarse y-position information was provided by the trigger scintillators, that the second detector is rotated by -1.8 mrad around the beam axis, which leads to the observed difference of the spatial resolution in pion and muon beams.

The spatial resolution, measured with muons and an Ar:CO$_2$ 93:7 vol% gas mixture, is displayed in Fig. 5.10. For correctly aligned detectors, it reaches a minimum value of $\sigma_{\text{SR}} = (30 \pm 2)$ µm for a low drift field of $E_{\text{drift}} = 0.3$ kV/cm. It should be noted, that the transverse electron diffusion, Fig. 2.5, is also minimal in this region. The spatial resolution degrades with increasing drift field, due to an increasing transverse diffusion. Although the transverse diffusion is constant for $E_{\text{drift}} \gtrsim 1$ kV/cm, the value of the spatial resolution increases for further increasing drift field.

In Fig. 5.11 the spatial resolution of the first Micromegas is shown as a function of the drift field, measured with and without rotation correction. The significant

Fig. 5.11 Spatial resolution of the first detector as a function of the drift field for $E_{amp} = 39.1\,kV/cm$ with and without correct rotation alignment of the four Micromegas. The first detector is rotated by $-1.8\,mrad$. Measured with muons using an Ar:CO$_2$ 93:7 vol% gas mixture. The influence of the additional y-hit information from the trigger scintillators and the thus possible rotation correction on a track-by-track basis is obvious

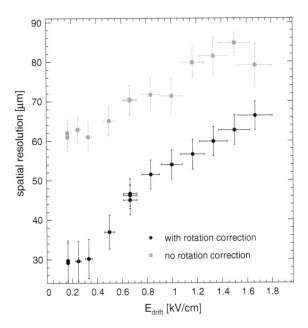

improvement of up to $\approx 30\,\mu m$ is visible, that is enabled by a coarse knowledge of the y-hit position

Generally speaking, the spatial resolution is, as above for the Ar:CO$_2$ 85:15 measurements, positively influenced by a small transverse diffusion in combination with a high detected ionization charge.

The difference in the absolute spatial resolution for pions and muons in an Ar:CO$_2$ 85:15 vol% gas mixture has been explained by relative rotation of the Micromegas. In the measurements with muons and the alternative 93:7 vol% gas mixture, the particle hit position could also be measured in the y-direction. This allows for the determination and thus the correction of relative rotation around the z-axis on an event-by-event basis (cf. Fig. 5.11).

5.1.5 Discharge Behavior

The discharge behavior in pion and muon beams has been studied in several dedicated runs. Discharges were registered by detecting the analog signal, created by the mesh recharge after a discharge. The particle rate is approximated by the coincident trigger rate of the two scintillator trigger layers.

In pion beams, similar discharge rates have been measured for all three operable Micromegas detectors. This is expected, when discharges are triggered by traversing particles. The discharge probability per incident pion as a function of the amplification field is shown in Fig. 5.12.

Fig. 5.12 Discharge
probabilities for the second
and the fourth Micromegas
as a function of the
amplification field, measured
at $E_{drift} = 0.8$ kV/cm with
pions at a flux density of
(12 ± 2) kHz/cm^2 using an
Ar:CO$_2$ 85:15 vol% gas
mixture. The errors are
calculated from the number
of discharges on the basis of
Poisson statistics. The
particle rate fluctuates by
about 10% during the
measurements

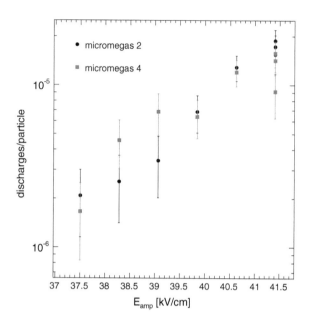

It increases exponentially with the amplification field in the observed range and
is on the order of 10^{-5} per particle for a pion flux density of (12 ± 2) kHz/cm^2. It is
shown in Sect. 6.2.7, that up to very high particle fluxes, the discharge probability is
independent of the particle hit rate.

The observed discharge rates in muon beams ranged between 2.3 h^{-1} for the first
and 13.0 h^{-1} for the fourth Micromegas for an observation time of 384 min at highest
amplification field. These discharges are probably not particle induced, but are due
to small detector defects. Nevertheless an upper limit for the discharge probability
per incident muon of $(9 \pm 3) \times 10^{-6}$ be calculated. This is similar to the discharge
probability for pions, but is most probably too high. Due to the relatively low muon
fluxes, no clear statement can be made about whether discharge probabilities in pion
and muon beams are qualitatively and quantitatively different. The discharge probabilities determined for pions, are expected to be a strong upper limit for discharge
probabilities in muon beams, due to the composite structure and thus additional
interaction mechanisms of pions.

5.1.6 Summary

A tracking telescope, consisting of four standard Micromegas, has been set up. It has
been tested with 120 GeV pions and 90 GeV muons at the SPS secondary beams test
area. Two different Ar:CO$_2$ based gas mixtures were used.

Charge signals from the Micromegas and hit information from the triggering scintillators were acquired with a Gassiplex based readout system. The system has been optimized with respect to trigger latency and readout rate.

The pulse height increases exponentially with the applied amplification field, as expected. No saturation effects are visible. The drift field dependence of the pulse height could be explained by an improving electron-ion-pair separation and a signal length decrease at low fields and increasing mesh opacity due to increasing transverse electron diffusion and changing electric field line configuration at higher fields.

The detection efficiency depends weakly on the amplification field. The dependence on the drift field is more pronounced. For $E_{drift} \gtrsim 0.4\,kV/cm$ an efficiency plateau with values above 0.95 is reached, for muons in Ar:CO_2 93:7 vol% efficiencies up to 0.99 are observed.

Optimum spatial resolutions on the order of $35\,\mu m$ have been reached at gas mixture specific drift fields, low transverse diffusion and high mesh transparency are correlated with a small value of the spatial resolution.

The particle rates in pion beams were sufficiently large to determine an amplification field dependent discharge probability per incident pion on the order of 10^{-5}. In muon beams only an upper bound of the same order could be determined. The dead time due to discharges is in pion beams on the order of a few percent and in muon beams negligible.

The calibration measurements have demonstrated the good suitability of Micromegas as reference detectors in high-energy pion and muon beams. A track resolution below $20\,\mu m$ in the center of the system is achieved. Both gas mixtures are suitable, due to the higher gas gain in Ar:CO_2 93:7 vol% at equal amplification field, this mixture is used in the subsequent measurements.

5.2 Large Floating Strip Micromegas in High-Energy Pion Beams

Test measurements with a $48 \times 50\,cm^2$ floating strip Micromegas in high-energy pion beams at the H6 beam line at SPS/CERN are described in this section. The homogeneity of the detector has been investigated with respect to pulse height, efficiency and spatial resolution. By rotating the Micromegas with respect to the beam, the single plane angular resolution has been determined for the so called µTPC reconstruction method. This method enables an alternative hit position definition. The angular dependence of the spatial resolution is shown. The particle induced discharge behavior is discussed. Excerpts from this section have been published by Bortfeldt et al. (2013a).

Due to the 30 mbar relative overpressure in the Micromegas system, the drift gap of the large floating strip detectors is considerably deformed. In a subsequent parasitic measurement with high-energy muons, the deformation could be measured spatially resolved.

In the sections, in which measurement results are described, the electric field dependence is discussed first, followed by a comparison of different irradiated positions in the floating strip Micromegas.

5.2.1 Setup

The setup, trigger- and readout-electronics, used for the floating strip Micromegas test measurements is shown in Fig. 5.13. The floating strip detector has been described in detail in Sect. 3.4.3. Its $48 \times 50\,cm^2$ active area is formed by 1920 copper anode strips, with 150 µm width and 250 µm pitch. Anode strips are individually connected via printed 10 MΩ resistors to high-voltage, the signal is decoupled over a second layer of readout strips below the anode strips. Charge signals were acquired with APV25 based front-end boards, read out using the Scalable Readout System, Sect. 4.1.2. The detector was mounted at three points in a stable steel frame, that allowed for adjusting the x-position in 20 mm steps and the rotation around the y-axis in 5° steps. The detector is operated at an elevated pressure to prevent oxygen contamination of the detector gas and improve operational stability. The readout structure and the cathode plane are reinforced by 10 mm thick aluminum plates, into which nine equally spaced $120 \times 120\,mm^2$ windows have been cut. In order to investigate the detector homogeneity, the detector has been irradiated in the two outer windows of the upper row and the middle window in the center row.

A telescope, consisting of six $9 \times 10\,cm^2$ standard Micromegas with 360 strips and two $9 \times 9\,cm^2$ resistive strip Micromegas with two-dimensional strip readout, was used as accurate track reference. The telescope is an improved version of the system, discussed in Sect. 5.1. The detector strips pointed in y-direction, perpendicular hit information was provided by the two additional strip layers of the resistive-strip detectors. All strips of the standard Micromegas were read out using the Gassiplex based system, discussed in Sect. 4.1.1. For the resistive strip detectors, APV25 based front-end electronics interfaced with the Scalable Readout System have been used.

Traversing pions were detected by two scintillator layers, consisting of three individual $33 \times 100 \times 10\,mm^3$ scintillators, read out with Hamamatsu R4124 photo-multipliers (Hamamatsu Photonics K.K. 2010). The trigger signal for both readout electronic systems was derived from a coincident hit in both layers. A VME CAEN V775N time-to-digital converter recorded the timing of the six individual trigger scintillators, that provided an additional coarse position information in y-direction.

Since both systems, and especially the Scalable Readout System, miss trigger signals, offline alignment of data streams was enabled by acquiring the global trigger number with each electronics separately: in the VME based Gassiplex system, a CAEN V775N time-to-digital converter was used to record the output of a custom 12 bit event counter, named Triggerbox, Sect. 4.1.3. A dedicated APV25 front-end board was used to record the attenuated logic signals with the Scalable Readout System. Since the Triggerbox output signals have a fixed delay with respect to the

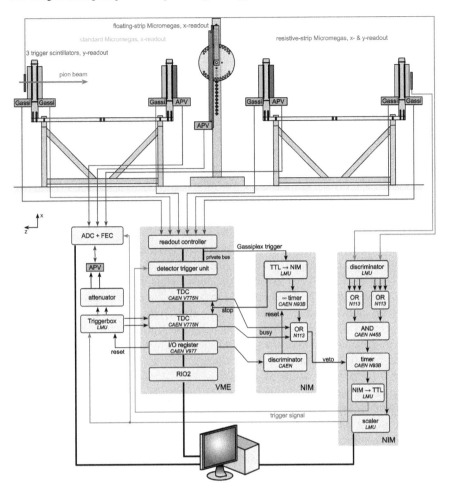

Fig. 5.13 Schematic setup of the floating strip Micromegas (*red*) and the track reference telescope (*green* and *magenta*). The eight reference detectors were mounted on two stable aluminum frames, reducing their relative rotation. Pions traverse the detector system in negative z-direction, their hit position in x-direction is precisely measured by all Micromegas. The triggering scintillators and the two perpendicular strip layers of the resistive strip Micromegas provide hit position information in y-direction. The y-axis points into the paper plane. In the *boxes* below are shown from *right* to *left* the trigger logic-, busy-, Gassiplex- and APV25-readout electronics

original trigger signal, this enabled furthermore the reduction of the 25 ns time jitter, encountered in the APV25 system, Sect. 4.1.2.

In order to avoid missed triggers as much as possible, the internal busy logic of the VME components, i.e. Gassiplex readout and TDCs, was used to generate a global busy signal (second to right block in Fig. 5.13).

The detectors were flushed with a premixed Ar:CO_2 93:7 vol% gas mixture with a flow of 2 ln/h. The gas pressure in the detectors was stabilized at (1013 ± 1) mbar using the gas system, described in Sect. 4.2. Due to the approximately 30 mbar over-pressure in the detector system, the drift gap of the large floating strip Micromegas is considerably deformed, Sect. 5.2.4.

Discharges between the mesh and the anode strips were counted for the floating strip Micromegas and three reference detectors with a FPGA based logic scaler, developed by the author. The FPGA controlled scaler reliably counts NIM signals with up to 80 MHz rate and communicates with the DAQ computer over an RS232 interface.[1] The high-voltage for the gas detectors was provided by several two-channel iseg SHQ 224M high voltage supplies (iseg Spezialelektronik GmbH 2012).

5.2.2 Pulse Height Behavior

In the following section, the measured pulse height as a function of the drift and amplification field is discussed. The distribution of pulse heights of the hits, selected in track building, is analyzed by fitting the pulse height distribution with a Landau function, convoluted with a Gaussian. The most probable pulse height is used for the following discussion. The dependence of the pulse height on drift and amplification field at a certain detector position is shown and then furthermore the homogeneity of the large floating strip Micromegas is investigated by comparing measured pulse heights at three different detector positions.

In Fig. 5.14 the pulse height as a function of the amplification field can be seen. The detector has been irradiated in three of the nine windows: the two outer windows of the upper row and the middle window in the center row, see Sect. 3.4.3. The pulse height, which is directly proportional to the gas gain, increases exponentially with increasing amplification field as expected [cf. Eq. (2.21)].

Comparing the pulse height, measured in different regions of the floating strip Micromegas, yields only small variations below 20%.

The pulse height as a function of the drift field for different amplification fields is shown in Fig. 5.15. For increasing drift field, the pulse height increases due to an improved separation of ionization charge, a reduced electron attachment to gas atoms and a decreasing signal length. Only the charge, arriving during the preamplifier integration time is detected. Maximum values are reached for $E_{\mathrm{drift}} \sim 0.4\,\mathrm{kV/cm}$. For further increasing drift field, the increasing transverse electron diffusion (Fig. 2.5) and the changing electric field configuration leads to a decrease of the measured pulse height, see also Sect. 2.6.

The good detector homogeneity is again visible in Fig. 5.16, where the drift field dependence of the pulse height is shown for three different irradiated positions in the detector. Variations below 20% are observed.

---

[1] The firmware has been written by A. Zibell, additional features like channel masking, gated counting, common-stop- and common-start-TDC capabilities were added by J. Ebke.

Fig. 5.14 Pulse height as a function of the amplification field at three different positions in the floating strip Micromegas. Measured with perpendicularly incident 120 GeV pions at $E_{drift} = 0.33$ kV/cm. An exponential function has been superimposed to guide the eye

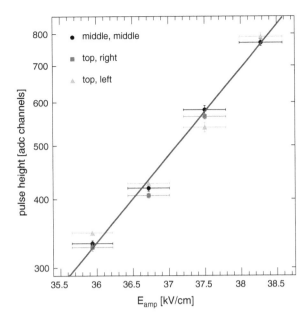

Fig. 5.15 Pulse height as a function of the drift field for different amplification fields, measured with perpendicularly incident 120 GeV pions, traversing the floating strip Micromegas in the upper right window

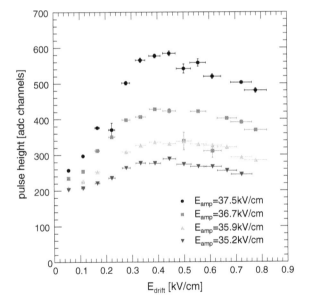

5.2.3 Efficiency

Two different quantities are used to characterize the detection efficiency of the floating strip Micromegas for 120 GeV pions: the hit efficiency and the track based

Fig. 5.16 Pulse height as a function of the drift field for three different beam positions. Measured with perpendicularly incident pions at $E_{amp} = 36.7\,kV/cm$. The two outliers at 0.50 and 0.61 kV/cm are caused by low statistics in the corresponding measurements, due to large blocks of missed triggers in the SRS, caused by DAQ computer lock-ups

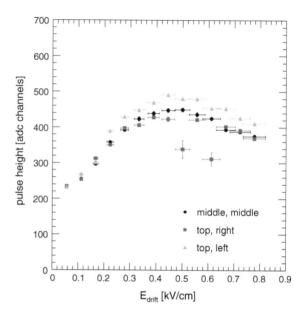

efficiency, Sect. 4.6. The hit efficiency for a certain detector is determined by selecting events in which all other detectors registered a hit and then testing, whether the detector under investigation also registered a hit.

For the track based efficiency, "good" tracks in the other detectors are interpolated into the detector under test. If a valid hit is found in a $\pm 0.3\,mm = 5 \cdot \overline{\sigma_{SR}}$ wide region around the expected position, the event is counted as efficient and as inefficient otherwise. $\overline{\sigma_{SR}} \approx 60\,\mu m$ is assumed as mean spatial resolution (cf. Fig. 5.28).

The accuracy of the determined efficiency values is calculated with Eq. (4.18), assuming Poisson statistics. If vertical error bars are not visible, the markers are larger than the errors. The efficiency has intentionally not been corrected for systematic effects due to discharges. It was rather intended, to demonstrate their influence and to present the actual efficiency, that is achievable in a realistic application.

In Fig. 5.17 the hit and the track based efficiency of the floating strip Micromegas are shown as a function of the drift field for different amplification fields. The reference track based efficiency is about 2 % lower than the simple hit based efficiency. Starting at small drift fields, the efficiency increases rapidly and reaches a plateau at $E_{drift} \gtrsim 0.2\,kV/cm$. Optimum values above 0.95 are reached at $E_{drift} \sim 0.4\,kV/cm$. The increasing mesh opacity at higher drift field then leads to a slight decrease of the efficiency. Since the amplification fields were chosen rather low, the plateau-efficiency shows a significant dependence on the amplification field. For later application of the large floating strip Micromegas, slightly higher amplification fields should be used.

The relatively strong fluctuation between the measured values for different drift fields in the amplification region is caused by two effects:

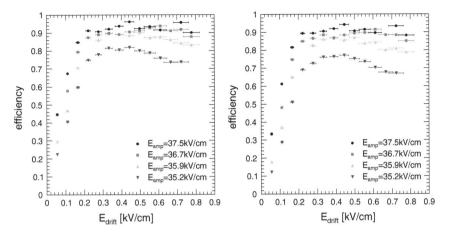

Fig. 5.17 Hit efficiency (*left*) and reference track based efficiency (*right*) as a function of the drift field for different amplification fields. Measured with perpendicularly incident 120 GeV pions, traversing the floating strip Micromegas in the upper right window

1. The 1920 anode strips are connected in eight groups of 7×256 strips and 1×128 strips to a common high-voltage distribution point. A $10\,\mathrm{M\Omega}$ series resistor between the high-voltage supply and each strip group limited the minimum recharge time, such that the dead time after a discharge was on the order of 350 ms. Furthermore the used iseg supplies are not designed for quick readjustment of the high-voltage when suddenly currents are drawn. This problem has been avoided in later measurements by relying on the individual strip resistor for current limitation and by introducing additional resistors as loads to the high-voltage supply, such that the recharge current after a discharge is by a factor of ten smaller than the current over the load resistor.

2. The Scalable Readout System was used in combination with the Gassiplex system for the first time in the presented measurements, leading to a limited data collection efficiency in the early measurements. Thus, a measurement for a single voltage configuration typically yields 5000 usable events, such that the efficiency is significantly influenced by the loss of events after a discharge. Due to the low discharge rate, only a few discharges per run are observed, such that the statistical fluctuation of the number of discharges leads to the observed efficiency fluctuation.

Part of the inefficiency for perpendicularly incident particles is due to the mesh supporting pillars, as can be seen in Fig. 5.18, where the inefficient regions in the floating strip Micromegas are shown. Clearly visible is the periodic rhombic pattern, that vanishes in measurements with inclined tracks. The vertical gap between $y = 32\,\mathrm{mm}$ and $y = 35\,\mathrm{mm}$ is caused by a 3 mm wide gap between the triggering scintillators. Pions, crossing the floating strip Micromegas in this region, were not detected by the trigger system.

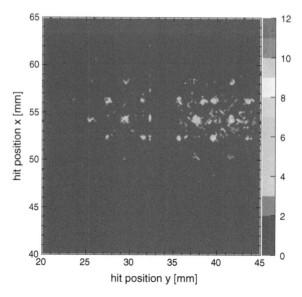

Fig. 5.18 Inefficient spots in the floating strip Micromegas, measured with perpendicularly incident pions at $E_{amp} = 36.7\,\text{kV/cm}$ and $E_{drift} = 0.33\,\text{kV/cm}$

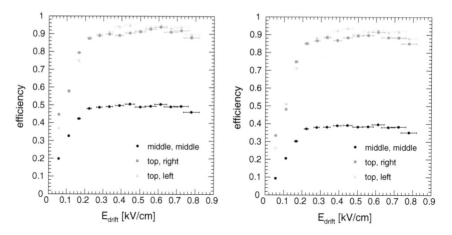

Fig. 5.19 Hit efficiency (*left*) and reference track based efficiency (*right*) as a function of the drift field for three different irradiated detector windows. Measured with perpendicularly incident 120 GeV pions at $E_{amp} = 36.7\,\text{kV/cm}$

The homogeneity of the efficiency over the detector can be extracted from Fig. 5.19, where the efficiency as a function of the drift field, measured at three different positions in the detector is shown. Except for the discussed fluctuation of the measured efficiency due to discharges, the extracted values for the two upper windows agree. For the middle window in the center row though, the observed efficiency is by a factor of 2 lower.

This can be understood as follows. The incoming pions hit strips that belong to two different high-voltage groups. A high-ohmic short circuit between an anode strip and the mesh in one of the two groups leads, in combination with the global $10\,M\Omega$ series resistor to a decrease of the common high-voltage by 50 V, rendering all strips in the group inefficient. The strips in the correctly functioning group are unaffected. For the later measurements with inclined detector, the high-voltage of the strip group incorporating the defective strip has been elevated by 50 V to compensate for the voltage drop. Removing the $10\,M\Omega$ series resistor, would also resolve the problem.

5.2.4 Pressure-Induced Drift Gap Deformation

All gas detectors were operated at about 30–40 mbar overpressure with respect to the surrounding. Due to the large active area of the floating strip Micromegas, this leads to a considerable enlargement of the drift region, which becomes visible especially in measurements with inclined detector. These measurements were performed to investigate the single plane angular reconstruction capabilities with the so called µTPC-reconstruction method (Sect. 4.8). The reconstructed track inclination depends crucially on the accurate knowledge of the electron drift velocity in the drift region. A variation of the drift gap leads to a variation of the electron drift velocity, Fig. 5.20. This can be measured directly by determining the maximum electron drift time (Sect. 4.8.6) or indirectly, using the µTPC-reconstruction method.

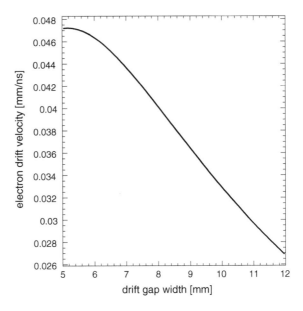

Fig. 5.20 Electron drift velocity as a function of the drift gap width for a drift voltage $U_d = 300\,V$, computed with MAGBOLTZ (Biagi 1999) for an Ar:CO$_2$ 93:7 vol% gas mixture at 1013 mbar

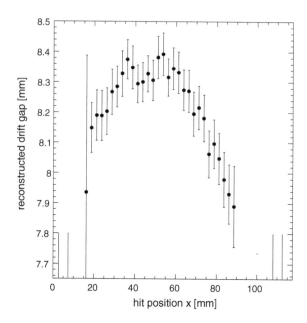

Fig. 5.21 Reconstructed drift gap width in the upper window of the floating strip Micromegas as a function of the predicted hit position, measured with high-energy muons at a detector inclination with respect to the beam of $(29 \pm 1)°$. The projected edges of the windows in the upper aluminum lid (*red*) and the lower aluminum lid (*blue*) are superimposed

In Fig. 5.21 the reconstructed drift gap width d_d in the floating strip Micromegas as a function of the hit position in x-direction is shown. It has been determined by measuring the maximum electron drift time t_{max} and then solving

$$v_d \left(E_{drift} = \frac{U_d}{d_d} \right) = \frac{d_d}{t_{max}} \tag{5.2}$$

for d_d. The drift velocity $v_d \left(\frac{U_d}{d_d} \right)$ as a function of the electric drift field U_d/d_d has been calculated with MAGBOLTZ. A measurement with high-energy muon beams has been used for this analysis, since the beam intensity is approximately constant over the investigated area of $9 \times 9\,cm^2$.

As we integrated over the hit position in y-direction, the shown values are lower bounds for the true drift gap width. This is especially important in measurements with a localized beam, hitting the detector in the middle of a window, such as in the inclined measurements, presented in Sect. 5.2.5, where even larger deformations are observed.

Deformations of the drift gap on the order of 0.5 mm are observed in this measurement. The FR4 material, carrying the cathode, and the readout structure are considerably deformed in the cutouts of the aluminum lid and base plate. But also a deformation of the aluminum plates themselves is visible. The latter can be seen from the difference between the reconstructed values for low and high x-hit positions, which do not seem to converge towards the same value.

Fig. 5.22 Measured signal timings, determined with the method, described in Sect. 4.3 for three different absolute pressures in the floating strip detector. Measured at $E_{amp} = 36.7\,kV/cm$ and $U_{drift} = 300\,V$

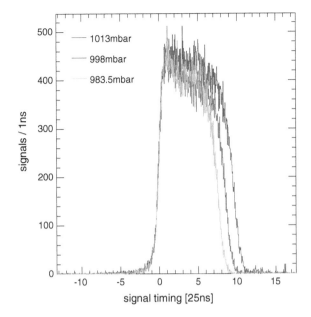

The electron drift velocity itself depends only weakly on the absolute pressure. As discussed above, the maximum electron drift time depends on the drift gap width and on the electron drift velocity, Eq. (5.2). The pressure in the floating strip Micromegas has been varied in measurements with a detector inclination of 29°, in order to determine the pressure induced inflation. The detector has been irradiated in the middle window of the center row. In Fig. 5.22 the reconstructed signal timing distributions are shown for three measurements at equal amplification field, equal drift voltage but different pressure.

A clear shift of the maximum electron drift time, defined by the upper edge of the spectrum is visible. This can be translated into a drift gap variation. In Fig. 5.23 the reconstructed drift gap width is shown, a variation of 1.1 mm is visible.

In direct measurements of the pressure induced inflation with a large caliber in the laboratory in Munich, pressure induced deformations on the order of 1.5 mm have been observed.

From these measurements we can conclude, that the detector is first considerably deformed due to the overpressure of the detector gas, and second, the deformation is rather inhomogeneous.

Due to small systematic uncertainties in the reconstruction of the drift gap width from the *absolute value* of the measured maximum electron drift time, an unbiased method is discussed in the following, that allows for a determination of the actual drift gap width in the inclination measurements with the floating strip detector.

In Fig. 5.24 the reconstructed electron drift velocities as a function of the drift field for different assumed drift gaps are shown for three different detector inclinations, see Sect. 4.8.6 for a discussion of the method.

Fig. 5.23 Reconstructed
drift gap width in the floating
strip Micromegas as a
function of the absolute gas
pressure, measured with
120 GeV pions at a detector
inclination with respect to
the beam of $(29 \pm 1)°$

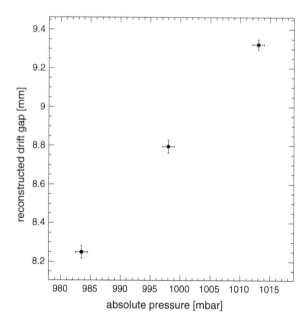

Comparing the shape of the determined curves with the expected velocities from a
MAGBOLTZ calculation, allows for determining the drift gap for the different measurements. A systematic mis-reconstruction of the electron drift velocity is observed,
that might be due to a slightly incorrect determination of the maximum drift time
or a deficiency of MAGBOLTZ. In any way is the sign of the deviation expected to
be constant. The drift gap can be approximated by searching for the drift velocity
curve, that agrees best with the expected values at medium drift fields and for which
the reconstructed velocities are always larger than the expected values.

The best agreement in the presented case is thus achieved for $d_d = 8.8$ mm for
the 20°, $d_d = 9.0$ mm for the 29° and $d_d = 9.5$ mm for the 38° measurements. Since
the floating strip Micromegas was not rotated about its center axis, a rotation leads
also to an overall shift of the detector, explaining the variation of the drift gap width
between the measurements.

5.2.5 Angular Resolution

The single plane track angle reconstruction capability of the floating strip Micromegas
is discussed in the following section. The method has been described in Sect. 4.8,
where the reconstruction method and systematic deviations due to capacitive coupling of neighboring strips and edge effects have been discussed in detail.

In order to investigate its angular reconstruction capabilities, the floating strip
Micromegas has been tilted with respect to the reference detectors about the y-axis,

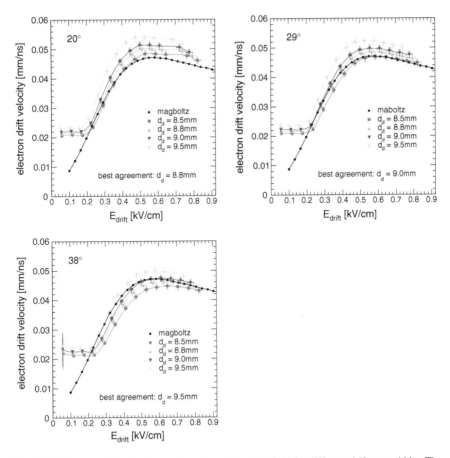

Fig. 5.24 Electron drift velocity as a function of the drift field for different drift gap widths. The values have been determined by measuring the maximum drift time. By comparing the shape of the measured velocities with values, calculated with MAGBOLTZ, the actual drift gap width can be determined. Measured with 120 GeV pions and a detector inclination of 20° (*top left*), 29° (*top right*) and 38° (*lower left*) at $E_{amp} = 37.5$ kV/cm

Fig. 5.13. Due to torsion of the pivot under the weight of the floating strip detector, the actual inclinations were $(10 \pm 1)°$, $(20 \pm 1)°$, $(29 \pm 1)°$ and $(38 \pm 1)°$. These values have been determined independently by aligning the floating strip detector via tracks, Sect. 4.7.4.

An inaccuracy arises from the considerable deformation of the drift gap, caused by the over-pressure in the detector, see Sect. 5.2.4. The actual drift gap width has been determined by comparing the measured electron drift velocities to the expected, Fig. 5.24. The applied method uses the maximum electron drift time in the drift region and is independent of the μTPC-reconstruction. Note that in principle the drift gap width can be directly determined by comparing the true and the reconstructed track inclination angles. This has been avoided in order to not bias the reconstructed track angles and show the necessity of an accurate and homogeneous drift gap. In an application, where the reconstructed angle is *used*, this calibration would nevertheless be performed.

The reconstructed track inclination angle and the angular resolution are extracted from the distribution of reconstructed track angles using a fit with a piecewise defined Gaussian function, Sect. 4.8.3. Since systematic effects such as capacitive coupling of neighboring anode strips and mis-interpreted noise lead to reconstructed track angles, that are larger than the true track angle, the distribution shows tails towards larger values. Thus the reconstruction accuracy for angles lower and higher than the most probable value are quoted separately.

In Fig. 5.25 the most probable reconstructed track angle and the angular resolution is shown as a function of the amplification field for the four different track inclinations. For the measurements under $10°$, the reconstruction method barely works, only for the highest amplification field, results are reliable.

At higher track inclinations the behavior is as follows: The reconstructed most probable angle approaches the true track inclination from above with increasing amplification field, the angular resolution improves accordingly. Best results and an accuracy of $\left(^{+4}_{-3}\right)^°$ are reached for the highest amplification field at the highest inclination. The observed amplification field dependence is caused by mis-identification of noise on neighboring strips as valid signal, which becomes less probable with increasing pulse height.

The most probable reconstructed angles and the angular reconstruction accuracy as a function of the drift field at different detector inclinations are shown in Fig. 5.26. Starting at low drift fields, the reconstructed angles decrease due to an increasing pulse height, which leads to an improving reconstruction. The reconstructed angles become minimal for $0.2\,\mathrm{kV/cm} \lesssim E_{\mathrm{drift}} \lesssim 0.3\,\mathrm{kV/cm}$ and even reach values smaller than the true inclination. This is caused by a strong dependence of the electron drift velocity on the electric field for $E_{\mathrm{drift}} \lesssim 0.4\,\mathrm{kV/cm}$ (Fig. 5.24), such that a small inaccuracy in the assumed drift gap width leads to a noticeable deviation of the reconstructed angle, also towards unphysical values. For $E_{\mathrm{drift}} \gtrsim 0.3\,\mathrm{kV/cm}$ the reconstructed values increase due to the leveling of the electron drift velocity, leading to a smaller dependence on the actual drift gap width. Furthermore the increasing mesh opacity leads to a decreasing pulse height, causing a progressive contribution by mis-interpreted noise signals.

A combination of relatively small electron drift velocity and high pulse height leads to optimum angular accuracies in the region $0.25\,\mathrm{kV/cm} \lesssim E_{\mathrm{drift}} \lesssim 0.35\,\mathrm{kV/cm}$. A quantitative comparison of the measured and the expected systematic deviations of the most probable reconstructed angle due to capacitive coupling of neighboring strips is given in Sect. 6.1.5 for a measurement with $20\,\mathrm{MeV}$ protons and a small floating strip Micromegas, where a good agreement between measured and expected values is shown.

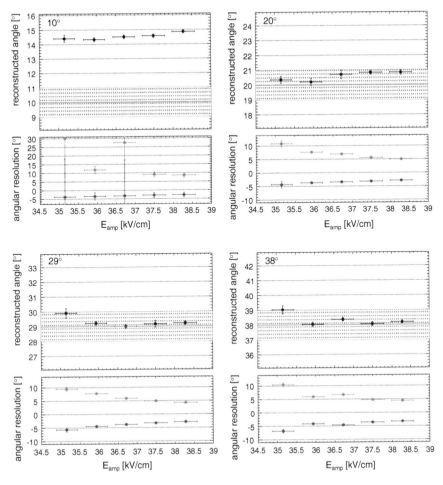

Fig. 5.25 Reconstructed track inclination and angular resolution as a function of the amplification field for true track angles of $(10 \pm 1)°$ (*top left*), $(20 \pm 1)°$ (*top right*), $(29 \pm 1)°$ (*lower left*) and $(38 \pm 1)°$ (*lower right*). Measured with 120 GeV pions at $E_{\text{drift}} = 0.3$ kV/cm. The angular accuracy for angles higher than the most probable value is represented by positive values and vice versa. Note the different scale for the angular resolution for $10°$ track inclination

5.2.6 Spatial Resolution

The spatial resolution of the floating strip Micromegas for perpendicular and for inclined tracks is discussed in the following section. It has been determined by interpolating reference tracks, measured with six reference Micromegas, into the floating strip detector. Its spatial resolution is then determined with the method, described in Sect. 4.5.3: The residual between the hit position measured in the floating strip Micromegas and the hit position predicted by the reference tracks is calculated

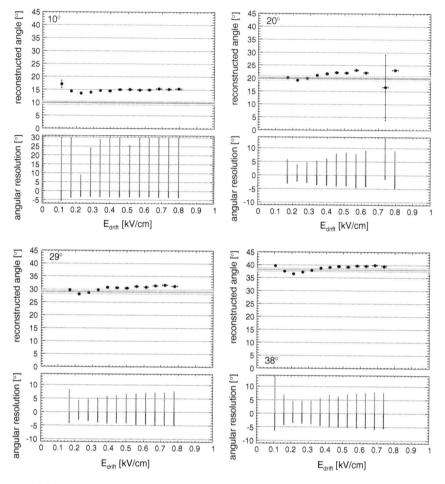

Fig. 5.26 Reconstructed track inclination and angular resolution as a function of the drift field for true track angles of $(10 \pm 1)°$ (*top left*), $(20 \pm 1)°$ (*top right*), $(29 \pm 1)°$ (*lower left*) and $(38 \pm 1)°$ (*lower right*). The angular accuracy for angles higher than the most probable value is represented by positive values and vice versa. Note the different scale for the angular resolution for $10°$ track inclination. The $10°$ measurements were performed with $E_{amp} = 36.7\,kV/cm$, the rest with $E_{amp} = 37.5\,kV/cm$

for all events, in which the reference track quality is sufficiently good.[2] The hit position measured in the floating strip detector is explicitly excluded from the track fit. The width of the residual distribution, determined from a fit with a double Gaussian function, and the track accuracy can then be used to calculate the spatial resolution of the floating strip detector with Eq. (4.16).

---

[2]The quality of the line fit to the hit positions in the reference detector is defined by $\chi^2/n_{ref.\,detectors}$. Good reference tracks are defined by $\chi^2/n_{ref.\,detectors} \leq 3$.

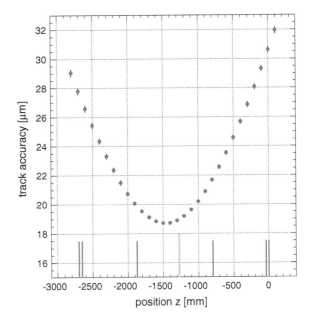

Fig. 5.27 Track accuracy as a function of the z-position, determined for measurements with 120 GeV pions, traversing the floating strip detector in the middle window of the center row. The position of the reference detectors is marked by *blue lines*, a *green line* is used for the floating strip detector. The asymmetry of the track accuracy is caused by a worse spatial resolution in the detector at $z = 0$ mm due to a relative rotation, that cannot be completely corrected for

The track accuracy has been calculated from the spatial resolution of the reference detectors with Eq. (B.11) and is exemplarily shown in Fig. 5.27. The reference detector spatial resolution has been determined with the geometric-mean method, Sect. 4.5.2, yielding 40–50 µm for the different reference detectors.

In Fig. 5.28, the spatial resolution of the floating strip Micromegas for perpendicularly incident particles is shown as a function of the drift field for different amplification fields. It improves with increasing amplification field, as the single strip signal becomes larger, leading to an improved definition of the hit position. Optimum values are reached for $0.2\,\text{kV/cm} \lesssim E_{\text{drift}} \lesssim 0.35\,\text{kV/cm}$, where also the transverse electron diffusion is minimal, see Fig. 2.5. Note, that the electron diffusion seems to have a larger impact on the spatial resolution than the overall pulse height, which is maximal at slightly higher drift fields $E_{\text{drift}} \sim 0.4\,\text{kV/cm}$, cf. Fig. 5.15.

In Fig. 5.29 the spatial resolution, measured at three different positions in the floating strip Micromegas is shown as a function of the drift field. The observed values agree within their respective uncertainties, showing that the spatial resolution is homogeneous over the detector.

Up to now, the hit position in the detector under test has been defined by the charge-weighted mean of all strips in a cluster, Eq. (4.2). This method yields robust and reliable results for perpendicularly incident tracks, but quickly degrades for inclined tracks, as can be seen in Fig. 5.30. Alternatively, the zero of the µTPC fit (Sects. 4.8 and 5.2.5) can be used as a measure of the hit position. For track inclination angles above 18°, this yields considerably better spatial resolutions σ_{SR} with only a small dependence on the track inclination. The drawback of the method is the lower

Fig. 5.28 Spatial resolution as a function of the drift field for different amplification fields, measured with perpendicularly incident 120 GeV pions, traversing the floating strip Micromegas in the upper right window

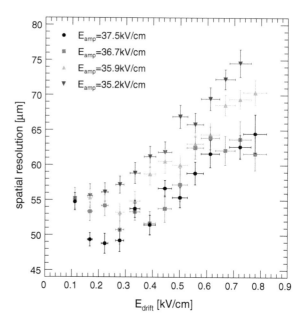

Fig. 5.29 Spatial resolution as a function of the drift field for three different irradiated detector windows. Measured with perpendicularly incident pions at $E_{amp} = 36.7$ kV/cm

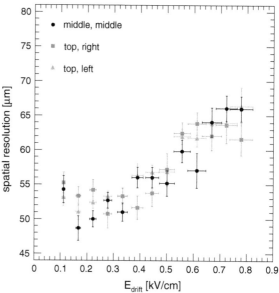

robustness, as the μTPC fit delivers reliable hit positions only for part of the triggered events. This is equally visible in the tails of the distribution of reconstructed angles. For track inclinations above 18°, $(64 \pm 1)\%$ of the used events lie within the $3\sigma_{SR}$ band around the central peak of the residual distribution, for track inclinations of 10°,

Fig. 5.30 Spatial resolution as a function of the track inclination angle, measured at $E_{amp} = 37.5\,\text{kV/cm}$ and $E_{drift} = 0.3\,\text{kV/cm}$. The particle hit position in the conventional method is defined by the charge-weighted mean of all strips in a hit cluster, alternatively it can be defined by the zero of the μTPC-fit

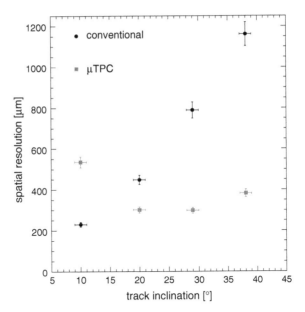

the fraction reduces to $(49 \pm 1)\%$. This can probably be increased by operating at higher amplification fields or, more laborious, using a preamplifier electronics with better temporal resolution.

5.2.7 Discharge Behavior

Discharges between mesh and anode strips of the floating strip Micromegas have been monitored by registering the discharge signal at the common high-voltage distributors via 100 pF capacitors. Due to the large coupling between detector strips, it was only possible to register the occurrence of a discharge, not to localize it. After amplification and shaping with an Ortec 452 Spectroscopy Amplifier and subsequent discrimination, the signals were counted in a FPGA based logic scaler.

By comparing the total discharge rate $\dot{n}_{dt}$ to the rate of traversing pions $\dot{n}_{\pi}$, discharge probabilities per incident pion could be calculated. The amplification field dependent rate of spontaneous discharges $\dot{n}_{ds}$, which are caused by small detector defects and thus do not depend on the incident pion rate, has been determined by exploiting the spill structure of the beam at the H6 beam line: Pions are delivered over an interval of about 8 s, followed by a break of about 40 s. Discharges registered during the spill break are assumed to be spontaneous.

The discharge probability per incident particle is then given by

$$P_p = \frac{n_{d\pi}}{n_\pi} = \frac{\dot{n}_{dt} - \dot{n}_{ds}}{\dot{n}_\pi} . \tag{5.3}$$

Fig. 5.31 Discharge probability per incident pion as a function of the amplification field for three different beam positions. Measured with perpendicularly incident pions at $E_{drift} = 0.33 \, kV/cm$

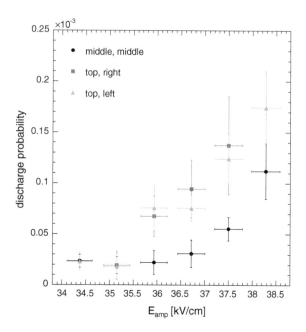

The measured discharge probabilities as a function of the amplification field and thus the gas gain are shown in Fig. 5.31. The pion flux density for the presented measurements was on the order of $(1.1 \pm 0.4) \, kHz/cm^2$. The discharge probability shows a strong dependence on the amplification field, as expected. The extracted behavior is equal for detector irradiation in two upper windows and by a factor of 2 smaller in the middle window of the center row. This is due to the reduced amplification field for approximately half of the irradiated strips in this region, see Sect. 5.2.3.

The measured pion induced discharge probability as a function of the drift field is shown in Fig. 5.32 for different amplification fields. At small drift fields, the discharge probability is highest, which can be explained by the low transverse electron diffusion in this field region. The critical charge density, necessary for the development of a streamer and a subsequent discharge, can thus be reached even for smaller ionization charge, see Sect. 2.7.

Assuming a discharge probability of 1.5×10^{-4}, we expect at a mean particle rate of $1.8 \pm 0.3 \, kHz$ a discharge rate of $(0.27 \pm 0.05) \, Hz$. Using the large dead time of the floating strip detector after a discharge of 350 ms, we thus expect a mean efficiency drop of $(9.5 \pm 2.0)\%$, which explains the observed maximum efficiencies well, cf. Fig. 5.17.

Note that the determined discharge probabilities are upper bounds for the true values, since the pion rate is in any case underestimated. As can be seen from Fig. 5.18, there exists a 3 mm gap between trigger scintillators, such that pions, crossing the floating strip Micromegas are missed by the trigger system.

Fig. 5.32 Discharge
probability per incident pion
as a function of the drift field
for different amplification
fields, measured with
perpendicularly incident
120 GeV pions, traversing
the floating strip Micromegas
in the upper right window

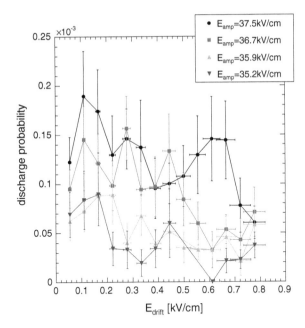

5.2.8 Summary

A floating strip Micromegas with an active area of $48 \times 50\,cm^2$ has been tested in
high-energy pion beams. Tracks were reliably measured with a Micromegas telescope, consisting of six standard Micromegas with one-dimensional readout and two
resistive strip Micromegas with two-dimensional readout. The track accuracy at the
position of the floating strip detector was 19.0 μm. Reliable offline merging of data
from the two applied readout systems was possible.

Due to the overpressure of approximately 30 mbar in the gas detectors, the drift gap
of the floating strip detector is inhomogeneously deformed by several millimeters.
The relative deformation has been measured directly by analyzing the maximum
electron drift time. The actual drift gap width has been determined by comparing the
drift field dependence of reconstructed and calculated electron drift velocities.

The pulse height response of the floating strip Micromegas is homogeneous with
relative variations below 20 %. An exponential dependence on the amplification field
is observed as expected, the dependence on the drift field shows the expected behavior.

The detection efficiency of the floating strip Micromegas is above 95 % for optimized electric fields, it is limited by discharges due to the chosen high-voltage powering scheme. The discharge probability per incident pion is of the order 1.5×10^{-4}.

Measurements with tilted detector allowed for an investigation of the single
plane track angle reconstruction capabilities of the floating strip Micromegas. The
reconstructed track inclinations are mainly larger than the true track inclination and
show the expected inclination dependent systematic behavior. For 10° inclination

a deviation of $+5°$ is observed, that reduces quickly for increasing track inclination. For optimized electric field parameters, the angular resolution is better than $5°$ for inclinations $\vartheta \geq 20°$.

An optimum spatial resolution of $(49 \pm 2)\,\mu m$ is reached with the floating strip Micromegas, no inhomogeneities are observed. For tracks with inclinations above $18°$, hit reconstruction based on the µTPC method yields up to a factor of three more accurate results than the usual charge-weighted mean reconstruction.

The measurements have shown, that the large floating strip Micromegas works in principle as expected. Three major changes are foreseen for future applications of this detector:

1. The supportive aluminum plates will be replaced by a light but stable FR4-aluminum honeycomb structure. If the operation at an overpressure on the order of 30 mbar is desired, the cathode layer should not simultaneously serve as gas-tight detector lid. An additional gas tight skin should be added.
2. The high-voltage powering scheme has to be optimized further along the lines, described in Sect. 5.2.3: The high-voltage should be buffered with large capacitors and a filtering circuit, to be independent of the high-voltage supply reaction time. Global series resistors in the common high-voltage line should be avoided. A new version of the detector will be equipped with larger strip recharge resistors on the order of $30\,M\Omega$.
3. Assembly and commissioning of additional detectors would be simplified by not attaching the micro-mesh to the readout structure permanently, but gluing the mesh to the aluminum gas frame, such that the amplification region can be opened and cleaned.

5.3 Cosmic Muon Tracking with a Floating Strip Micromegas in High-Rate Proton Background

In the following, the detection capabilities of a floating strip Micromegas for minimum ionizing particle are investigated in a high-rate background environment at the tandem accelerator in Garching. A small floating strip Micromegas detector with 128 strips is laterally irradiated with a high-rate 20 MeV proton beam, that crosses the whole active area, such that a single proton can deposit charge on all anode strips. The proton rate was about 550 kHz such that on average in every third event proton *and* cosmic muons signals are present. Such an event is shown in Fig. 5.33, where the earlier muon signal is clearly distinguishable from the later proton signals.

Four non-irradiated reference Micromegas are used to detect cosmic muons. Interpolating the muon track into the irradiated Micromegas under test, allows for an investigation of the muon detection efficiency, pulse height and spatial resolution. Due to the small muon trigger rate of $4\,min^{-1}$ only two electric field parameter points, $E_{amp} = 35.6$ and $36.3\,kV/cm$ at a constant drift field $E_{drift} = 0.27\,kV/cm$, have been investigated under proton irradiation.

Fig. 5.33 Typical event in the irradiated floating strip Micromegas, measured with $E_{amp} = 38.7\,kV/cm$ and $E_{drift} = 0.27\,kV/cm$. A cosmic muon signal is clearly visible on strips 59–61 at $t = 10 \times 25\,ns$, the traversing proton creates coincident signals on many strips at $t = 20 \times 25\,ns$

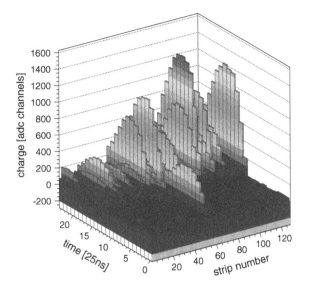

After completion of the irradiation measurements, cosmic muon measurements were performed without proton background with the same setup. This allows for a direct comparison of results and for unfolding detector and background effects.

5.3.1 Setup

The schematic detector setup, together with trigger, busy and readout electronics can be found in Fig. 5.34.

The tested floating strip Micromegas with an active area of $6.4 \times 6.4\,cm^2$ and 128 anode strips, with $300\,\mu m$ width and $500\,\mu m$ pitch, has been described in detail in Sect. 3.4.1. The anode strips are coupled via discrete HV capacitors to the APV25 based readout electronics, described in Sect. 4.1.2. The traversing protons are registered and counted by a scintillator detector. By defocussing and wobbling in horizontal y-direction, a beam focus of $60 \times 2\,mm^2$ has been realized.

Cosmic muon tracks are measured with two standard, non-floating strip Micromegas with an active area of $9 \times 10\,cm^2$, subdivided into 360 strips, with $150\,\mu m$ width and $250\,\mu m$ pitch, and two resistive strip Micromegas with a $9 \times 9\,cm^2$ active area. Their readout structure is formed by two perpendicular layers of 358 readout strips with $150\,\mu m$ width. Strips of all five Micromegas point into the paper plane, i.e. the y-direction and measure particle hit position in the x-direction, Fig. 5.34. The strips of the second layer in the two resistive strip Micromegas point into x-direction and provide perpendicular hit position information.

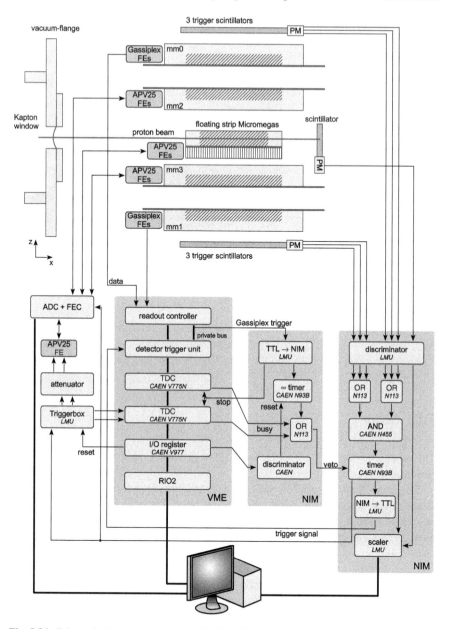

Fig. 5.34 Schematic detector setup, trigger logic and readout electronics. The active area of each detector is diagonally hatched. 20 MeV protons cross the active area of the floating strip Micromegas and are stopped in a scintillator. The readout systems are triggered by coincident muon signals in two scintillator layers. The scintillator hit information is acquired with a VME based TDC (connection not drawn). The *lower right block* shows the NIM based trigger logic, the *lower middle block* displays the NIM based busy logic, the VME readout system is shown in the *lower left block*. mm0 and mm1 are standard Micromegas with one-dimensional strip readout, mm2 and mm3 are two-dimensional resistive strip Micromegas

The two standard Micromegas are read out with the VME-based Gassiplex readout electronics (Sect. 4.1.1). APV25 based front-end boards, together with the Scalable Readout System (Sect. 4.1.2), are used to acquire time-resolved charge signals of the floating strip and the two resistive strip Micromegas. The combined readout system, busy logic and data stream merging is described in detail in Sect. 5.2.1.

Both readout systems are triggered by coincident cosmic muon hits in two scintillator layers, above and below the gas detectors. Each layer consists of three $33 \times 100 \times 10 \, mm^3$ scintillators, read out with small R4124 Hamamatsu photomultipliers (Hamamatsu Photonics K.K. 2010).

High voltage was provided and monitored by two 12 channel CAEN A1821 modules (CAEN S.p.A. 2013a) in a CAEN SY5527 mainframe (CAEN S.p.A. 2013b). Currents drawn by the detectors, were monitored with an accuracy of $(2 \, \%rdg \pm 10 \, nA)$.[3]

Discharges of the floating strip Micromegas, muon trigger signals and proton signals in the counting scintillator were counted with a FPGA based NIM counter, developed by the author.

The gas detectors were connected in series to the gas system, described in Sect. 4.2. Premixed Ar:CO_2 93:7 vol% gas at an absolute pressure of (1000 ± 2) mbar was constantly flushed through the detectors with a flow rate of 2 ln/h.

5.3.2 Gas Gain and Mesh Transparency

In the following section, the signals created in the irradiated floating strip Micromegas by traversing 20 MeV protons are discussed. Due to their large energy loss, a direct measurement of the current between mesh and anode strips allows for a determination of gas gain (Sect. 2.4) and electron mesh transparency behavior (Sect. 2.6).

A SRIM (Ziegler et al. 2010) based TRIM calculation has been used to determine the mean energy loss of initially 20 MeV protons in the active area of the irradiated detector, after passage through the Kapton beam window, air and the Kapton window in the detector frame. The mean energy loss in an Ar:CO_2 93:7 vol% gas mixture of $\Delta E / \Delta x = (3.08 \pm 0.01) \, keV/mm$ corresponds to an ionization yield of $q_0 = (7417 \pm 40) \, e/6.4 \, cm$, using the mean energy per created electron-ion-pair of 26.56 eV, given in Table 2.1. Due to the proton track length of 6.4 cm, the low proton energy and the measurement method, the mean, rather than the most probable energy loss, can be used in the discussion.

According to Eq. (2.25), the measured current on the high-voltage supply

$$I(f, E_{drift}, E_{amp}) = q(E_{drift})G(E_{amp})f \qquad (5.4)$$

---

[3]rdg: reading.

Fig. 5.35 Current between
mesh and anode strips as a
function of the proton rate
for different drift fields at an
amplification field
$E_{amp} = (35.0 \pm 0.7)\,kV/cm$.
The measured currents have
been fitted with a *line*
through the origin

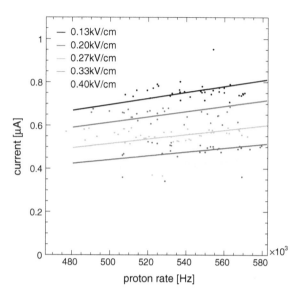

Fig. 5.35 Current between
mesh and anode strips as a
function of the proton rate
for different drift fields at an
amplification field
$E_{amp} = (35.0 \pm 0.7)\,kV/cm$.
The measured currents have
been fitted with a *line*
through the origin

depends on the gas gain $G(E_{amp})$, the particle rate f and the charge, reaching the
amplification region $q(E_{drift}) = q_0 t(E_{drift})$, which is given by the ionization yield
q_0 and the drift field dependent electron mesh transparency $t(E_{drift})$. The weak
dependence of the transparency on the amplification field (Kuger 2013) is neglected
in this discussion.

The measured currents are shown in Fig. 5.35 as a function of the proton rate for
different drift fields. Fitting the measured currents with a straight line through the
origin, i.e. assuming no rate dependence of the single particle pulse height, allows
for a determination of the gas gain and transparency. The slope p_0 of the fit function
can then be related via Eq. (5.4) to transparency and gas gain

$$p_0(E_{drift}) = q_0 t(E_{drift}) G(E_{amp}) . \qquad (5.5)$$

A GARFIELD simulation of the mesh transparency (Sect. 2.6) predicts $t(E_{drift} =
0.13\,kV/cm) = (0.82 \pm 0.02)$. Relying on the simulation, $p_0(E_{drift})/p_0(0.13$
$kV/cm)$ can be solved for the mesh transparency at arbitrary drift field, yielding

$$t(E_{drift}) = \frac{p_0(E_{drift}) t(0.13\,kV/cm)}{p_0(0.13\,kV/cm)} . \qquad (5.6)$$

The resulting measured transparency is shown in Fig. 5.36 and compared to the
simulated values, showing a very good agreement for all considered drift fields.

The measured currents as a function of the proton rate for different amplification
fields are shown in Fig. 5.37. Using the same method as above, the gas gain can be
calculated from the slope $p_0(E_{amp})$ of the fit function and the ionization yield q_0

Fig. 5.36 Electron mesh transparency as a function of the drift field in the floating strip Micromegas. Measured from traversing 20 MeV protons at an amplification field $E_{amp} = (35.0 \pm 0.7)$ kV/cm. The simulated transparency values, taken from Fig. 2.8, have been determined using a GARFIELD simulation

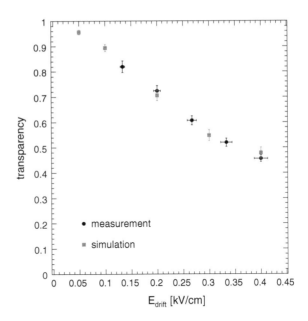

Fig. 5.37 Current between mesh and anode strips as a function of the proton rate for different amplification fields and a drift field $E_{drift} = 0.27$ kV/cm. The measured currents have been fitted with a *line* through origin

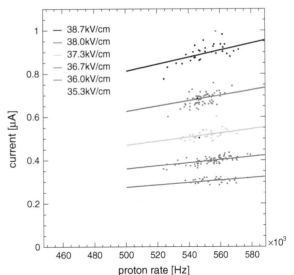

$$G(E_{amp}) = \frac{p_0(E_{amp})}{q_0 t(E_{drift})} , \tag{5.7}$$

using the measured value of $t(0.27 \text{ kV/cm}) = (0.59 \pm 0.02)$.

In Fig. 5.38, the resulting gas gain is displayed, showing the expected exponential behavior. The large errors of the amplification fields are due to a lacking knowledge

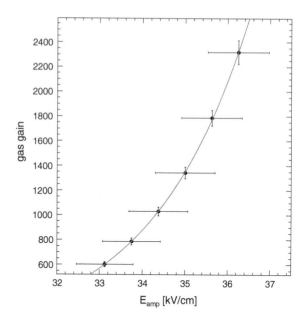

Fig. 5.38 Gas gain as a function of the amplification field in the floating strip Micromegas. Measured from traversing 20 MeV protons at a drift field $E_{drift} = 0.27$ kV/cm. An exponential function has been superimposed to guide the eye

of the exact width of the amplification gap. It will be shown in Sect. 6.1.3, that due to an initially imperfect production process, a slight swelling of the mesh supporting pillars on the order of 10 µm is observed. For this detector an amplification gap width of (160 ± 3) µm is assumed.

5.3.3 Proton Occupancy and Rate

The probability of reconstructing a proton signal, coincident with the triggering cosmic muon in the floating strip Micromegas, is discussed in the following section. The proton rate, measured with the proton counting scintillator and with the irradiated Micromegas, is discussed.

A proton typically produces a (350 ± 50) ns long signal, considering the involved capacitances and the shaper behavior of the APV25 readout chip. Cosmic muon signals are approximately 400 ns long, including the electron drift and the decay time of the shaper circuit. Upon a trigger signal from the scintillators, charge signals are acquired over a time window of 600 ns. The probability for detecting a proton signal in an event, triggered by a traversing cosmic, can be calculated from the Poisson distribution, see Appendix B.4. Since proton and cosmic muon hits are uncorrelated and protons, hitting the Micromegas up to 350 ns before the beginning of the acquisition window, are still detected, we can assume an effective time window of $T = (950 \pm 50)$ ns. Using Eq. (B.15) and a mean proton hit rate of 550 kHz, the

Fig. 5.39 Number of reconstructed clusters per event in a dedicated run, in which the proton counting scintillator provided the trigger signal. Measured at a proton rate of $(553 \pm 1)\,\text{kHz}$ with $E_{amp} = 35.6\,\text{kV/cm}$ and $E_{drift} = 0.27\,\text{kV/cm}$. In 99.4 % of the events, at least eight clusters are detected

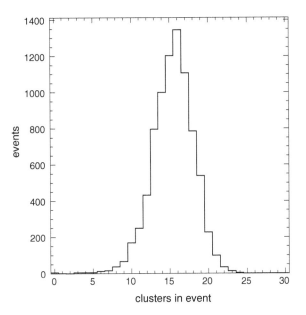

probability of finding at least one proton in an event, triggered by a traversing cosmic muon is $P = (0.38 \pm 0.02)$.

Dedicated runs were taken, where the readout electronics has been triggered by a signal from the proton counting scintillator. Due to the high proton rate, this corresponds to a sample, containing exclusively proton signals. This allows for determining the Micromegas proton detection efficiency and, furthermore, the contamination of the proton beam with X- and γ-rays and neutrons. These would artificially increase the measured "proton" rate, without depositing charge in the Micromegas.

In Fig. 5.39 the number of clusters in a dedicated, proton-triggered run is shown. A traversing proton creates on average 15 clusters in the Micromegas. Since in only 5 of 8153 events, not a single cluster is detected, the proton beam contamination with photons and neutrons is negligible and the Micromegas proton detection efficiency is very close to 1.

Since in 99.4 % of all events, a proton creates at least eight clusters in the Micromegas, the number of clusters can be used to identify cosmic muon events, in which an additional proton has been detected. Note that a cluster denotes in this context a group of neighboring hit strips. A coincident cosmic muon and a proton create thus at least nine clusters in the detector.

The number of reconstructed clusters in cosmic muon measurements with and without proton irradiation is shown in Fig. 5.40. The observed ratio of events including a proton signal can be determined by integrating the distribution over the number of clusters in an event. Events with less than nine clusters correspond to events in which only a cosmic muon has been detected, in events with more than nine clusters a proton has additionally been detected.

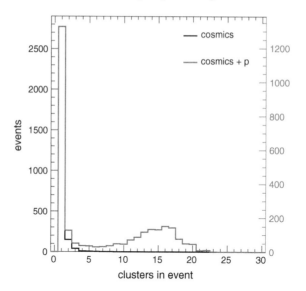

Fig. 5.40 Number of reconstructed clusters per event, measured at $E_{amp} = 35.6\,$kV/cm and $E_{drift} = 0.27\,$kV/cm. Shown are the clusters per event for muon tracking measurements with and without proton irradiation. In events with nine or more clusters, a traversing proton has clearly been detected together with the cosmic muon. For less then nine detected clusters, only a muon has been found. A fluctuating amplifier base line, which is due to proton charge deposit clearly before the beginning of the acquisition window, leads to the increased number of clusters per event with respect to the non-irradiated case

The ratio of events with a detected traversing proton is then

$$P_{\mu+p} = 1 - \frac{n_{\mu\,\text{only}}}{n_{\text{total}}} . \tag{5.8}$$

For measurements with the amplification field $E_{amp} = 35.6\,$kV/cm this yields a measured ratio of $P_{\mu+p} = (0.40 \pm 0.01)$.

When using a higher amplification field of $E_{amp} = 36.3\,$kV/cm, the number of clusters per event in pure proton measurements is smaller than before, see Fig. 5.41. This is due to the higher gas gain, leading to a merging of neighboring clusters. A proton creates at least five clusters per event.

Using again the number of charge clusters per event in measurements with cosmic muons and additional proton irradiation Fig. 5.42, the ratio of events with a reconstructed muon and at least one proton can be calculated from Eq. (5.8). This yields a ratio of $P_{\mu+p} = (0.39 \pm 0.02)$ for the higher amplification field $E_{amp} = 36.3\,$kV/cm.

The measured ratios of events, in which a traversing proton has been found additionally to the triggering cosmic muon, agree with each other and with the expected value for both amplification fields used.

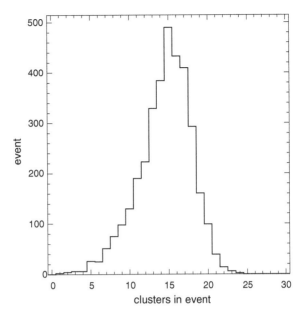

Fig. 5.41 Number of reconstructed clusters per event in a dedicated run, in which the proton counting scintillator provided the trigger signal. Measured at a proton rate of (553 ± 1) kHz with $E_{amp} = 36.3$ kV/cm and $E_{drift} = 0.27$ kV/cm. In 99.5 % of events, the proton creates at least five clusters

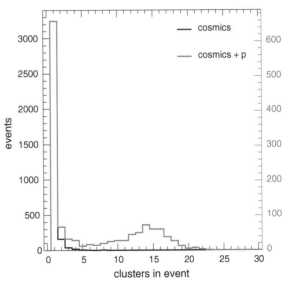

Fig. 5.42 Number of reconstructed clusters per event, measured at $E_{amp} = 36.3$ kV/cm and $E_{drift} = 0.27$ kV/cm. In events with six or more clusters, a traversing proton has clearly been detected together with the cosmic muon. For less then six detected clusters, only a muon has been found

In Fig. 5.43, the proton rate during the three-day irradiation is shown as a function of time. It has been measured with the proton counting scintillator. The rate is subject to variations on the order of only 20 %, showing the long time stability of the tandem accelerator.

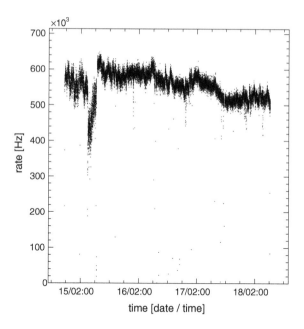

Fig. 5.43 Proton rate versus time, measured with the proton counting scintillator during the irradiation

5.3.4 Muon Track Reconstruction

The following section is intended for readers, interested in the specific hit and track reconstruction methods, that are used for cosmic muon reconstruction. It will be discussed, whether the track building algorithm biases the spatial resolution and efficiency measurements.

Cluster selection and reconstruction on detector level has been described in Sect. 4.3. In this particular experiment, the pulse height standard deviation, used to determine a hit strip, is increased in the floating strip Micromegas under irradiation, as small and out-of-time proton signals are interpreted as electronic noise. To avoid an artificial masking of small signals, the standard deviation is not calculated for every run, but is set to a constant and uniform value for all strips. The chosen value has been determined in the cosmic muon measurements without irradiation.

Cosmic muon tracks are reconstructed with the chain algorithm, described in Sect. 4.1.1. That means particularly, that in the irradiated floating strip Micromegas, the cluster is selected, that is closest to the hit position, predicted by the four non-irradiated reference detectors. If in a ±2 mm region around the hit prediction no valid cluster is found, the event is counted as inefficient. In order to not bias the measured spatial resolution of the floating strip Micromegas, the detector is not included in the fit, used to select the correct cluster combination.

Additional discrimination of proton and cosmic muon signals could in principle be achieved by using the measured signal timing. The cosmic muon signal is coincident in all tracking detectors, proton signals are uncorrelated. To be able to control the muon selection capabilities of the reconstruction method with an unbiased quantity, the signal timing selection is explicitly not used.

In the analysis only those events are considered, in which all five detectors registered a hit. Tracks, defined by the reference detectors, are used for the following analysis, if they intersect with the previously known active area of the floating strip Micromegas. Its active area of $6.4 \times 6.4 \, \text{cm}^2$ is considerably smaller than the $9 \times 9 \, \text{cm}^2$ coverage of the reference detectors. In order to improve the track accuracy, the inclination of the used tracks is restricted to almost perpendicularly incident muons with $-0.1 \, \text{rad} \leq \vartheta \leq 0.1 \, \text{rad}$.

The floating strip Micromegas spatial resolution can then be extracted from the distribution of residuals, i.e. of the difference between measured and predicted hit position.

In order to investigate, whether the chain algorithm breeds an artificial peak around 0 in the residual distribution of the irradiated detector, the assumed position of this detector is shifted in software by $1.0 \, \text{mm}$, Fig. 5.44. This is found not to be the case. Reconstructed cosmic muons form a narrow peak in the distribution, sitting on a flat background of proton and noise signals.

The residual distribution is fitted with the sum of a Gaussian function and a constant, Fig. 5.45,

$$f(\Delta x) = p_0 \exp\left(-0.5 \left(\frac{\Delta x - p_1}{\sigma_{\text{ex}}}\right)^2\right) + p_3, \tag{5.9}$$

Fig. 5.44 Residual distribution, measured with cosmic muons under proton irradiation. The position of the investigated detector has been shifted by $+1.0 \, \text{mm}$

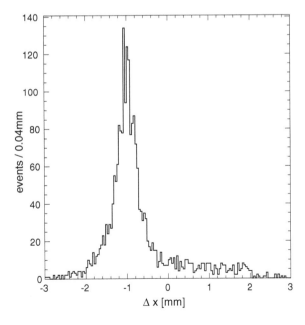

Fig. 5.45 Residual distribution, measured with cosmic muons under high-rate proton irradiation at $E_{amp} = 35.6\,$kV/cm and $E_{drift} = 0.27\,$kV/cm. The distribution has been fitted with a Gaussian function with offset (*red*)

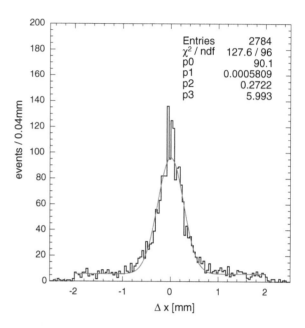

where the standard deviation, given by σ_{ex}, allows for the calculation of the spatial resolution. The proton contamination can be estimated from the pedestal p_3.

5.3.5 Spatial Resolution

The reference track accuracy at the position of the irradiated floating strip Micromegas of $\sigma_{track} = (50.4 \pm 0.5)\,\mu$m, has been determined with the geometric mean method (Sect. 4.5.2). This allows for the calculation of the spatial resolution, σ_{SR}, of the floating strip Micromegas, from the width of the residual distribution, σ_{ex}, with Eq. (4.16).

The residual distributions in cosmic muon measurements without proton irradiation are fitted with a simple Gaussian function without offset. Under irradiation, the procedure is as follows: the number of clusters per event in the irradiated detector is used as indicator for the coincident detection of a proton with the triggering cosmic muon. For $E_{amp} = 35.6\,$kV/cm, more than eight, for $E_{amp} = 36.3\,$kV/cm more than five clusters in an event signal the presence of an additional proton, compare Sect. 5.3.3.

The residual distributions are treated differently for events with only a cosmic muon and those with a cosmic muon together with a traversing proton. For events, in which only the triggering muon was detected, the distribution is fitted with a simple Gaussian function as above. A Gaussian function with offset Eq. (5.9) is used to fit the distributions for the other case. The offset represents the proton contamination

Fig. 5.46 Spatial resolution for cosmic muons as a function of the amplification field. Measured in the floating strip Micromegas with (*red squares*) and without additional proton irradiation (*black points*) at $E_{drift} = 0.27\,kV/cm$. The measurement with proton background is further separated into events with only a cosmic muon (*green triangles*) and with a cosmic muon and a coincident proton (*blue triangles*)

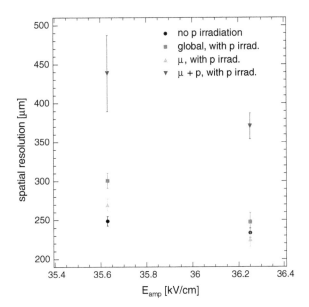

i.e. the number of proton or noise signals, that are incorrectly interpreted as cosmic muon signal by the track finding algorithm. The mean spatial resolution, i.e. for *all* events, is then determined by fitting the corresponding distribution with a Gaussian with offset, with the offset set to the fixed value, determined before.

The measured spatial resolution of the floating strip Micromegas is shown in Fig. 5.46. Without proton irradiation, an optimum value of $\sigma_{SR} = (233 \pm 5)\,\mu m$ is observed. It has improved by about 10 % for an increase from $E_{amp} = 35.6\,kV/cm$ to $E_{amp} = 36.3\,kV/cm$. The mean spatial resolution degrades slightly under proton irradiation, for events with only a cosmic muon, the spatial resolution is very similar as for the non-irradiated situation. The spatial resolution is in events with a coincident proton by about a factor of two worse, than for the measurement without background. This is due to a distortion of the measured cosmic muon hit position by additional proton signals on the same or on neighboring strips (Fig. 5.47).

It is remarkable, that traversing protons only significantly influence the spatial resolution for cosmic muons, if the same strip region is hit at the same time as by the muon. The Micromegas does not suffer from readout structure charge-up or discharge effects. The chosen setup, where a traversing proton creates a signal on all strips, corresponds to the worst possible background hit situation. It is clear, that in a strip detector, coincident charge depositions by two particles on the same strips cannot be separated. At the moment, protons that traverse the Micromegas approximately 300 ns before or after the passage of the cosmic muons, degrade the muon hit position reconstruction. By using a faster front-end chip and improving electron as well as ion drift velocity, this time window can be further reduced.

Fig. 5.47 Residual distribution versus number of clusters per event, measured with cosmic muons under proton irradiation at $E_{amp} = 35.6$ kV/cm and $E_{drift} = 0.27$ kV/cm. Clearly visible is the qualitative difference between the narrow distribution in events without a proton (few clusters per event), and with a proton (nine or more clusters per event)

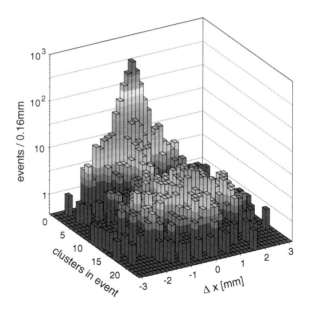

5.3.6 Muon Efficiency

In the following, the muon detection efficiency of the floating strip Micromegas detector with and without proton irradiation is discussed.

It is determined by selecting tracks, that are well reconstructed in the four reference detectors, see Sect. 5.3.4, and then searching for reconstructed hits with a maximum distance of 2.0 mm to the predicted hit position in the floating strip detector.

Since the expected hit position is known, inefficient spots can thus be identified in the detector. Under proton irradiation, proton signals, that are accidentally within the expected region, fake muon signals in events, in which the detector was actually inefficient. As discussed in Sect. 5.3.4, these fake hits are visible in the residual distribution as uniformly distributed pedestal. Their number can be calculated from the offset of the Gaussian fit function Eq. (5.9), see also Sect. 5.3.5.

In Fig. 5.48, the inefficient spots of the floating strip Micromegas, measured with cosmic muons without proton irradiation, are shown. The floating strip Micromegas active area has been superimposed. Clearly visible are the sharp edges of the active area on the top, bottom and on the right. The inefficient region on the left, i.e. towards small x, can be correlated to a group of 30 strips, that are read out over coupling capacitors of a different type, than the rest of the detector. Obviously is their capacitance smaller; this might either be a manufacturing fault, or is due to a different dielectric, leading to a stronger decrease of the capacitance with increasing strip voltage.

Fig. 5.48 Inefficient spots in the floating strip Micromegas, measured with cosmic muons without proton irradiation at $E_{amp} = 35.6\,kV/cm$ and $E_{amp} = 36.3\,kV/cm$ with $E_{drift} = 0.27\,kV/cm$. The active area of the detector under test has been superimposed (*red square*)

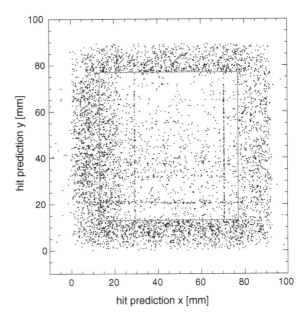

The two distinct inefficient vertical lines are caused by two groups of dead readout strips. The slightly thicker inefficient horizontal line at $y = 20\,mm$ is most probably due to a ripple in the micro-mesh, that is only glued to the aluminum gas frame.

The overall cosmic muon detection efficiency without background irradiation is thus (0.825 ± 0.010) for the lower and (0.875 ± 0.010) for the higher amplification field, Fig. 5.49. It significantly decreases to values around 0.62 under proton irradiation. Under irradiation, the contamination by accidental proton hits is subtracted by scaling the determined efficiency *with* proton contamination ε_{cont} with a correction factor. The corrected muon efficiency is then

$$\varepsilon = \varepsilon_{cont} \frac{n_{tot} - n_p}{n_{tot}}, \tag{5.10}$$

where n_p is the number of misidentified protons and n_{tot} the total number of efficient events. The number of protons can be calculated from the pedestal p_3 of the residual fit $n_p = p_3 n_{bins}$, where n_{bins} is the number of bins in the residual histogram between -2 and $2\,mm$ [Fig. 5.45, Eq. (5.9) and Sect. 5.3.5]. The total number of efficient events, the number of misidentified protons the corresponding efficiency correction factors are summarized in Table 5.1.

Assuming that a traversing proton blinds the detector completely, a muon efficiency under irradiation of 0.51 and 0.54 is expected. We can conclude that the muon detection efficiency is significantly degraded in events with a cosmic muon and a coincident proton, which is closely related to the corresponding degradation of the spatial resolution. The muon efficiency is most probably not significantly influenced by the irradiation, if no proton is coincidentally traversing the detector, as this would also adulterate the spatial resolution.

Fig. 5.49 Muon detection efficiency of the floating strip detector as a function of the amplification field. Measured with cosmic muon with and without additional proton irradiation and $E_{\text{drift}} = 0.27\,\text{kV/cm}$

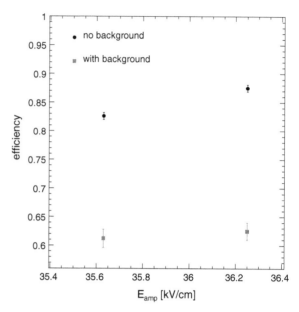

Table 5.1 Total number of efficient events n_{tot}, number of misidentified protons n_p and efficiency correction factors

E_{amp} (kV/cm)	n_{tot}	n_p	$(n_{\text{tot}} - n_p)/n_{\text{tot}}$
36.3	1076	166 ± 7	0.846 ± 0.007
35.6	2264	307 ± 29	0.864 ± 0.013

5.3.7 Discharge Behavior

Discharges between the mesh and anode strips are produced by large charge deposits from incident protons and are also due to small detector defects, that lead to spontaneous, uncorrelated discharges.

Discharges have been registered by detecting the recharge of anode strips at the common strip high-voltage distributor. The recharge signal was decoupled via a 1 nF capacitor, and was counted after amplification and discrimination with a 16 channel logic counter, compare Sect. 5.3.1. The same scaler has been used to measure the incident proton rate $\dot{n}_p = n_p/t$, by counting signals from the proton counting scintillator during the measurement time t.

Measurements without proton irradiation were used to determine the spontaneous discharge rate $\dot{n}_{ds} = n_{ds}/t$. Equation 5.3 allows then for a calculation of the proton induced discharge probability P_p. In Table 5.2 the observed probabilities and discharge rates are summarized. The spontaneous discharge rate $n_{ds}/t = (5.0 \pm 0.5) \times 10^{-3}\,\text{s}^{-1}$, used for the calculation of the discharge probability per incident proton, has been estimated from the measurements without proton irradiation.

Table 5.2 Total discharge rates n_{dt}/t in measurements with and without proton irradiation and discharge probability per incident proton P_p

E_{amp} (kV/cm)	Proton irradiation	n_{dt}/t (s^{-1})	P_p
36.3	Yes	$(173 \pm 2) \times 10^{-3}$	$(33.0 \pm 0.3) \times 10^{-8}$
35.6	Yes	$(54.2 \pm 0.6) \times 10^{-3}$	$(8.84 \pm 0.13) \times 10^{-8}$
36.3	No	$(4.28 \pm 0.09) \times 10^{-3}$	–
35.6	No	$(5.42 \pm 0.14) \times 10^{-3}$	–

Shown are the statistical errors according to Poisson, additional error contributions due to temperature variations etc. are not included

There is a significant increase of the discharge rate under proton irradiation, with respect to the measurements without proton background. Although the gas gain increases by only 28 % for an increase of the amplification field from $E_{amp} = 35.6$–36.3 kV/cm, (Fig. 5.38), the corresponding discharge probability increases by a factor 3.7. This is due to the threshold nature of discharge formation, as it has also been discussed in Sect. 2.7 and by Moll (2013). Note, that the observed discharge probabilities are rather low. Using the single strip dead time of about 1 ms and the fact, that at maximum three strips are affected, the overall inefficiency for the higher amplification field is $\cancel{\epsilon} = n_{dt}/t \times 1\,\text{ms} \cdot 3/128 = 4.1 \times 10^{-6}$ and thus negligible.

5.3.8 Summary

Cosmic muon tracking has been studied in a floating strip Micromegas under intense lateral irradiation with 20 MeV protons. A reference system, consisting of two standard and two resistive strip Micromegas provided reliable muon track prediction in the irradiated detector. Reference measurements without proton irradiation allow for unfolding irradiation and detector effects with respect to efficiency, spatial resolution and discharge behavior.

A proton is traversing the floating strip Micromegas coincidently with the triggering cosmic muon in $(38 \pm 2)\%$ of all events. The observed proton coincidence rate agrees well with the expectation. The proton rate on the order of 550 kHz has been measured with a scintillator detector.

The gas gain has been determined by measuring the proton induced current between micro-mesh and anode structure. The electron mesh transparency has been measured with the same method and agrees well with predictions from a GARFIELD simulation. No rate induced pulse height variations have been observed.

The floating strip Micromegas shows a spatial resolution of $\sigma_{SR} = (233 \pm 5)\,\mu\text{m}$ for cosmic muons without proton irradiation. Under proton background irradiation the mean spatial resolution degrades slightly to $\sigma_{SR} = (249 \pm 7)\,\mu\text{m}$. For events, in which only a cosmic muon has been detected, the spatial resolution is unchanged as compared to the case without proton background. In events, in which the cosmic muon is detected together with a transversing proton, the spatial resolution degrades to $\sigma_{SR} = (370 \pm 20)\,\mu m$.

A detection efficiency of (0.875 ± 0.010) has been observed for cosmic muons with proton background, the detection efficiency for traversing protons is very close to 1. The overall efficiency is limited by a group of 30 less efficient strips, that are coupled to the readout via low-quality capacitors. Under irradiation, the muon detection efficiency decreases to (0.625 ± 0.020). If the detector would be completely blinded by a traversing proton, an efficiency of 0.54 under irradiation is expected.

The discharge rate under proton irradiation is $0.173 \pm 0.002\,Hz$, equivalent to a discharge probability per incident proton of $(3.30 \pm 0.03) \times 10^{-7}$. Discharges have no influence on the detection efficiency.

The measurements show that the performance of the floating strip Micromegas is influenced by the background irradiation only in events, in which a cosmic muon traverses the Micromegas in coincidence with a proton. Even then, cosmic muon signals can be partially identified, tracking is still possible. It should be noted, that the presented lateral irradiation scheme represents the worst possible background situation, as a traversing proton can create coincident signals on all readout strips. In a strip detector, coincident hits by different particles on the same strip group can obviously not be separated. A Micromegas with a pixelized readout structure would perform even better in high-rate background.

5.4 Ion Backflow and Aging Measurements with a Resistive Strip Micromegas

In the following, ion backflow measurements in a resistive strip Micromegas are discussed. Ions produced in gas amplification processes drift from the anode towards the mesh, where they are neutralized. A small fraction can pass through the mesh into the drift region and drift slowly towards the cathode. This so called ion backflow can limit the efficiency and pulse height at ultra-high particle rates or in applications with very large drift gaps such as in Time-Projection-Chambers with Micromegas readout planes. The ion backflow has been measured in a resistive strip Micromegas under irradiation with 20 MeV protons at the tandem accelerator in Garching.

As the diffusion of positive ions is small, they are expected to follow the electric field lines. Thus, ions can only enter the drift region, if they are produced on field lines, that extent up to the cathode. Due to the large ratio between amplification and drift field, the drift field lines, that extent into the amplification region are strongly compressed, the ion backflow is expected to be small and field ratio dependent (Colas et al. 2004).

A 12 h irradiation with 20 MeV protons at increased rates has been used to deposit $0.43\,C/1.2\,cm^2$ in the detector. Charge deposition and sputtering aging effects can thus be studied. The behavior of the resistive strip Micromegas after the aging irradiation is currently under investigation (Danger 2014) and will not be treated in this thesis.

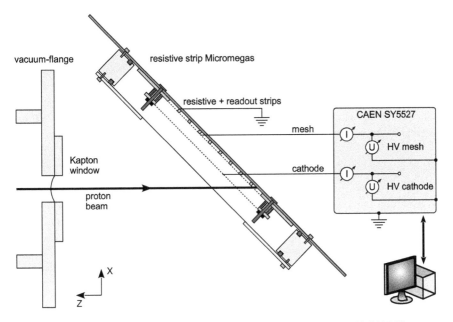

Fig. 5.50 Lateral view of the setup during the ion backflow measurements with 20 MeV protons. The currents drawn by the micro-mesh and the cathode are monitored over the high-voltage supply

5.4.1 Setup

The setup during the ion backflow measurements at the tandem accelerator is shown schematically in Fig. 5.50. An inclined resistive strip Micromegas with an active area of $9 \times 9 \, \text{cm}^2$ is irradiated with 20 MeV protons at a rate on the order of 6 MHz. The protons traverse the active volume and are stopped in the readout structure. The 358 resistive strips with a pitch of 250 μm and a width of 150 μm are oriented in vertical direction, the detectors features two perpendicular layers of mutually isolated readout strips below the resistive strip layer. The 358 strips per layer have a pitch of 250 μm. The resistive as well as the anode strips were grounded during the measurements. The 128 μm wide amplification gap is closed by a plain woven mesh with 400 lines per inch, consisting of stainless steel wires with 18 μm diameter. A 5 mm wide drift and ionization region is formed by the mesh and a cathode, consisting of the same material. The gas volume of the detector is closed by a 50 μm thick Kapton window.

Negative high-voltage for the mesh and the cathode is provided by a 12 channel CAEN A1821 module (CAEN S.p.A. 2013a) in a CAEN SY5527 mainframe (CAEN S.p.A. 2013b). Currents drawn by the mesh and the cathode can be monitored with an accuracy of $(2 \, \% \text{rdg.} \pm 10 \, \text{nA})$[4] with the high-voltage supply.

The DC proton beam creates a constant current of ionization charge. Electrons drift to the mesh and are amplified in the high-field region between mesh and anode.

---

[4]rdg: reading.

Positive ions from ionization processes drift accordingly to the cathode. This ionization charge current is below 1 nA and thus too small to be measured with the high-
voltage supply. This has been experimentally confirmed. The current of ionization
charge can thus be neglected in the discussion. Positive ions from gas amplification
drift from the anode to the mesh, where they are neutralized. A small fraction crosses
the micro-mesh and drifts towards the cathode.

By comparing the measured currents I_m and I_c, the relative ion backflow IB can
directly be determined

$$IB = \frac{I_c}{I_m + I_c} \approx \frac{I_c}{I_m} . \tag{5.11}$$

Note that the actual proton hit rate, the proton ionization yield and the gas gain
of the resistive strip Micromegas do not influence the measured IB, as long as the
currents are sufficiently large to be measured. This leads to a powerful method for
investigating the drift field dependence of ion backflow.

The Micromegas was flushed with Ar:CO$_2$ 93:7 vol% at atmospheric pressure
and a flux of 2 ln/h.

5.4.2 Ion Back Flow Measurement

The backflow of positive ions into the drift region is related to the transparency of the
mesh to electrons from ionization processes. A larger electron mesh transparency is
correlated to a smaller ion backflow and vice versa as both depend on the electric
field configuration.

The current between anode and mesh as a function of the drift field is shown
in Fig. 5.51 (left). It shows the typical Micromegas behavior: Starting at low fields,
the current increases due to an improving separation of ionization charge and lower
attachment. A maximum value is reached at $E_{drift} \sim 0.3$ kV/cm, due to a combination
of low attachment and low electron mesh opacity. For further increasing drift field, the
observed current decreases due to an increasing mesh opacity, caused by increasing
transverse electron diffusion and changing electric field configuration, Sect. 2.6.

In Fig. 5.51 (right), the measured ion backflow is shown as a function of the
drift field. It increases with increasing drift field due to the changing electric field
configuration. For typical drift fields on the order of $E_{drift} \sim 0.3$ kV/cm, the ion
backflow reaches 1%. Values on the order of 0.2% can be reached for even lower
fields. This small value demonstrates the good high-rate capability of Micromegas
detectors as compared to wire chambers, in which the produced ions have to drift
for several milliseconds towards the cathode planes. In Micromegas, 99% of the
produced ions are neutralized after a $\sim$150 ns drift towards the mesh.

Note that the shown backflow values are insensitive to fluctuations of the particle
rate as well as gas gain variations due to e.g. detector charge up. In case the amplification field close to the mesh is considerably altered by charge-up of the resistive anode

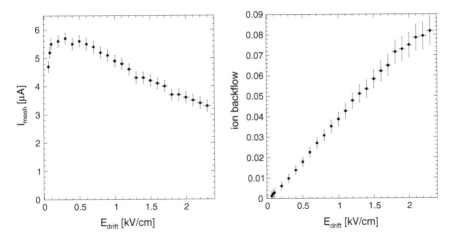

Fig. 5.51 Current between anode and mesh (*left*) and relative ion backflow (*right*) as a function of the drift field, measured with 20 MeV protons with an approximate rate of 6 MHz at an amplification field of $E_{amp} = 42.2$ kV/cm

structure, an effect on the ion backflow is expected, since the backflow depends on the ratio of drift and amplification field. This would also lead to a strong reduction of the electron mesh transparency, which agrees with the expectation (see e.g. Kuger 2013), obtained for negligible ion backflow.

5.4.3 Summary

The ion backflow has been measured in a resistive strip Micromegas under intense irradiation with 20 MeV protons. Comparison of the currents on micro-mesh and cathode allow for a determination of the ion backflow, that is independent of the particle rate and anode charge-up effects. The ion backflow increases approximately linearly from 0.002 at low drift fields to 0.08 at $E_{drift} = 2.3$ kV/cm.

References

Biagi SF (1999) Monte Carlo simulation of electron drift and diffusion in counting gases under the influence of electric and magnetic fields. Nucl Instrum Methods A 421(12):234–240. doi:10.1016/S0168-9002(98)01233-9. http://www.sciencedirect.com/science/article/pii/S0168900298012339; ISSN 0168-9002
Bortfeldt J (2010) Development of micro-pattern gaseous detectors—micromegas. Diploma thesis, Ludwig-Maximilians-Universität München. http://www.etp.physik.uni-muenchen.de/dokumente/thesis/dipl_bortfeldt.pdf

Bortfeldt J, Biebel O, Hertenberger R, Lösel Ph, Moll S, Zibell A (2013a) Large-area floating strip micromegas. PoS EPS-HEP2013:061

Bortfeldt J, Biebel O, Hertenberger R, Ruschke A, Tyler N, Zibell A (2013b) High-resolution micromegas telescope for pion- and muon-tracking. Nucl Instrum Methods A 718(0):406–408. doi:10.1016/j.nima.2012.08.070. http://www.sciencedirect.com/science/article/pii/S0168900212009734; ISSN 0168-9002 (Proceedings of the 12th Pisa Meeting on Advanced Detectors)

CAEN S.p.A. (2013a) Technical Information Manual Mod. A1821 Ser Rev 3. http://www.caen.it

CAEN S.p.A. (2013b) SY5527—SY5527LC power supply system, 7th edn. http://www.caen.it

Colas P, Giomataris I, Lepeltier V (2004) Ion backflow in the micromegas TPC for the future linear collider. Nucl Instrum Methods A 535(12):226–230. doi:10.1016/j.nima.2004.07.274. http://www.sciencedirect.com/science/article/pii/S0168900204016080; ISSN 0168-9002 (Proceedings of the 10th International Vienna Conference on Instrumentation)

Danger H (2014) Diploma thesis (in preparation)

Hamamatsu Photonics K.K. (2010) Photomultiplier tube R4124. https://www.hamamatsu.com/

iseg Spezialelektronik GmbH (2012) Bedienungsanleitung für Präzisions-Hochspannungs-Netzgeräte der Baureihe SHQ-HIGH-PRECISION. http://www.iseg-hv.com

Kuger F (2013) Simulationsstudien und Messungen zu Gasverstärkungsprozessen in Micromegas für den Einsatz im ATLAS NewSmallWheel. Master's thesis, Julius-Maximilians-Universität Würzburg

Moll S (2013) Entladungsstudien an Micromegas-Teilchendetektoren. Diploma thesis, Ludwig-Maximilians-Universität München

Ziegler JF, Ziegler MD, Biersack JP (2010) SRIM—the stopping and range of ions in matter. Nucl Instrum Methods B 268(1112):1818–1823. doi:10.1016/j.nimb.2010.02.091.http://www.sciencedirect.com/science/article/pii/S0168583X10001862; ISSN 0168-583X (19th International Conference on Ion Beam Analysis)

Chapter 6
Specific Applications of Floating Strip Micromegas

Floating strip Micromegas are intended for ion tracking in medical imaging applications. In the following chapter, the application of floating strip Micromegas as tracking detectors in high-intensity, low and medium energy ion beams, is presented.

A proof of concept study with 20 MeV proton beams demonstrated the low energy ion tracking capabilities of a floating strip Micromegas doublet with low material budget. The measurement allowed for a characterization of the detectors and a considerable improvement of reconstruction methods. The study and the results are presented in Sect. 6.1.

In ion beam therapy (Sect. 1.3.1), accurate prediction of the ion beam parameters is essential. Accurate imaging before and in between irradiation cycles could improve the treatment effectiveness and quality. The achievable resolution in ion radiography and tomography (Sect. 1.3.2) is expected to improve considerably, if single ion tracking is possible. The suitability and the limitations of floating strip Micromegas are investigated in beam characterization measurements at the Heidelberg Ion Therapy center. The measurements with very high-intensity beam enabled furthermore a stringent test of the high-rate capability of floating strip detectors. The measurements are presented in Sect. 6.2.

Detailed studies of the underlying functional principles of floating strip Micromegas have already been presented in Chap. 5. In this chapter, results are occasionally only summarized, measurement discussion is moved to the appendix, where appropriate.

6.1 Floating Strip Micromegas in High-Rate 20 MeV Proton Beams

In the following, proof of concept test measurements with a floating strip Micromegas doublet with low material budget in high-intensity 20 MeV proton beams at the tandem accelerator Garching are presented.

© Springer International Publishing Switzerland 2015
J. Bortfeldt, *The Floating Strip Micromegas Detector*, Springer Theses,
DOI 10.1007/978-3-319-18893-5_6

Measurements with perpendicularly incident beam were conducted with particle rates between 440 Hz and 475 kHz. The pulse height behavior and the detection efficiency has been studied. The determined spatial resolution is strongly distorted by the beam divergence and by multiple scattering. The rate behavior has been studied for different particle rates. During the high-rate irradiation, the gas gain could be deduced by measuring the current between mesh and anode structure. Due to the short measurement time of only a few minutes for each voltage setting and the low discharge rate, discharge probabilities in the proton beam could be determined only approximately.

The TPC-like single plane track angle reconstruction capabilities were investigated by tilting the detector doublet with respect to the beam by $10°$, $20°$, $30°$ and $40°$ at particle rates around 500 kHz. Electron drift velocities could be determined from an inversion of the μTPC reconstruction, the proton ionization yield was measured and compared to the expectation.

In the sections, in which measurement results are discussed, we focus first on the electric field dependence, measured in low rate, perpendicularly incident proton beams. The behavior with inclined detectors is discussed afterwards.

6.1.1 Setup

The Micromegas doublet, specifically designed for low-energy proton tracking, has been described in detail in Sect. 3.4.2. Two floating strip Micromegas with an active area of $6.4 \times 6.4 \, cm^2$, with 128 strips per layer of 300 μm width and 500 μm pitch, are mounted back to back, forming a single unit. A mesh of 25 μm wires is employed, rather than the 18 μm wire meshes, used for e.g. the $48 \times 50 \, cm^2$ floating strip Micromegas. In order to ensure a homogeneous amplification gap, the readout structures were pressed against the meshes by applying ∼8 mbar overpressure between the two readout structures. A thin steel cannula in the gas circuit between the space between the two readout structures and the active volume of the detectors is used to dynamically produce the overpressure. As the mesh tension is sufficiently high, no deformation of the 6.0 mm wide drift region, as defined by the detector frame, is observed.

The setup, the trigger circuit and the readout electronics circuit is shown schematically in Fig. 6.1. Charge signals from the Micromegas were acquired using one 128 channel APV25 front-end board per layer, read out with the Scalable Readout System (Sect. 4.1.2). Traversing protons are detected and stopped in a 1 cm thick $9 \times 9 \, cm^2$ scintillator, read out with a Hamamatsu R4124 photomultiplier (Hamamatsu 2010), that provided the trigger signal for the readout system. By counting the scintillator trigger signals, the proton rate could be measured. Since no coincident triggering was possible, the photon and neutron contamination of the beam was carefully checked. Typically more than 98 % of all trigger signals corresponded to a traversing particle.

The first i.e. the upstream detector layer is called layer 0 or Micromegas 0 in the following. The second layer is accordingly denoted with layer 1 or Micromegas 1.

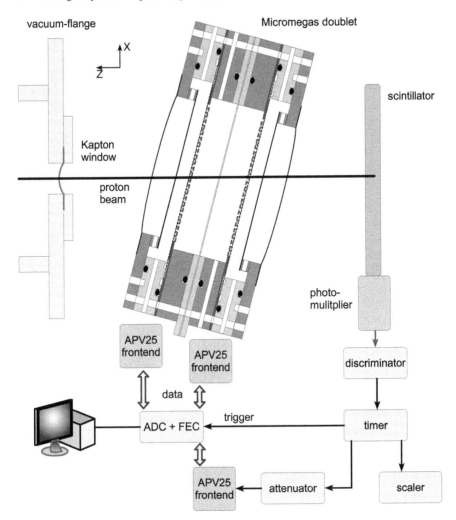

Fig. 6.1 Schematic setup in the floating strip Micromegas test measurements with 20 MeV protons at the tandem accelerator Garching. A Micromegas doublet with $6.4 \times 6.4 \, cm^2$ active area and 128 readout strips per layer is read out by one APV25 front-end board per layer, using the Scalable Readout System. Traversing protons are stopped in the $9 \times 9 \times 1 \, cm^3$ scintillator, providing the trigger signal for the Micromegas readout. The attenuated trigger signals are acquired with an additional APV25 front-end board to enable correction of the 25 ns time jitter of the Micromegas signals. Protons are moving into negative z-direction, their hit position in the x-direction is measured by the micromegas, in y-direction, that points into the drawing plane, no hit information is available

Discharges were monitored by detecting the analog recharge signal at the common high-voltage distribution for each Micromegas. The recharge signals were amplified in Ortec 452 Spectroscopy Amplifiers, discriminated, elongated to 3 μs and counted in a FPGA based 16 channel high-rate capable scaler.

High voltage was provided and monitored by two 12 channel CAEN A1821
modules (CAEN S.p.A 2013a) in a CAEN SY5527 mainframe (CAEN S.p.A 2013b).
Currents drawn by the detectors, can be monitored with an accuracy of (2 %rdg $\pm$
10 nA).[1]

The detectors were constantly flushed with an Ar:CO_2 93:7 vol% gas mixture at
atmospheric pressure with a flow of 2.0 ln/h. See Sect. 4.2 for a description of the gas
system. The atmospheric pressure during the measurements[2] was (965 $\pm$ 3) mbar.

6.1.2 Particle Rate and Flux Determination

The particle rate has been measured by counting the trigger scintillator signals. At
the employed maximum particle rates of 550 kHz, the dead time of the scintillator
detector is assumed to be negligible, the rate should be underestimated by less than
10 %. In Fig. 6.2, the measured rates are shown as a function of time.

Whereas the rate can be measured quite accurately, the flux density is not easily
determinable. The beam size at the position of the detectors has been measured by
replacing the detectors with a bare scintillator tile, and observing the produced light
spot with a camera. For the low rate measurements at perpendicular beam incidence, a
roughly rectangular beam spot with width (22.4 $\pm$ 0.5) mm in horizontal y-direction
and height (1.3 $\pm$ 0.2) mm in vertical x-direction was set up. For the high-rate
measurements and the angular scans, an elliptic beam spot of (4.5 $\pm$ 0.2) $\times$ (3.5 $\pm$
0.2) mm^2 in y- and x-direction has been used.

Due to stability requirements of the tandem accelerator, a minimum beam current
of several nA had to be used, such that up to three absorbers on the high-energy side
of the machine had to be used to reach the desired particle rates. Due to the sieve-like
structure of the absorbers, the particle flux density became highly inhomogeneous,
although the outer beam dimensions, measured with the scintillator tile, are expected
to remain unaffected. The beam spot measurement with the scintillator tile is only
possible when extracting all absorbers from the beam line. It should be noted, that
in principle the beam spot can of course be measured in situ by the Micromegas. For
the presented test beam though, only two thin Micromegas were available, such that
only hit information in vertical x-direction could be recorded.

For the calculation of the particle flux, we assume a homogeneous beam spot,
which leads in any case to an underestimation of the flux density.

6.1.3 Pulse Height Behavior

In the following section, the behavior of the pulse height of signals of 20 MeV protons
is discussed as a function of the amplification and drift field for perpendicularly

---

[1]rdg: reading.

[2]taken from: http://www.meteo.physik.uni-muenchen.de/.

Fig. 6.2 Measured proton
rate during the test beam
from 8th to 10th August
2013. The measurements
with perpendicular beam
incidence were conducted on
8/8 and 9/8 around 14:00,
runs with inclined detectors
were taken afterwards. A
dedicated set of
measurements at high gas
gain and low rate completed
the test beam on 9/8

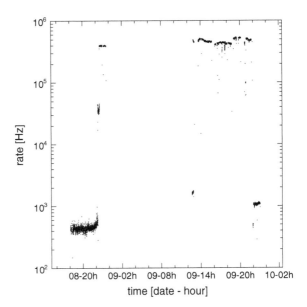

incident and for inclined tracks. The influence of the particle rate on the pulse height
is shown and the absolute gas gain is deduced.

By fitting the measured pulse height distributions with a Landau function, convoluted with a Gaussian function, the most probable pulse height has been determined.
Its value is used in the following analysis of the pulse height behavior. The fit error is
used as accuracy for the determined pulse height values in the figures in this section.
If no error bars are visible, the markers are larger than the respective errors.

For high particle rates, the currents I between the mesh and the anode strips are
on the order of several 10 nA and can be used to determine the absolute gas gain
(Eqs. 5.4 and 5.7). Since 20 MeV protons lose several MeV in the anode PCBs, the
energy loss in the two detector layers is different. It has been calculated with a TRIM
calculation (Ziegler et al. 2010) assuming a simplified setup, consisting of the 50 μm
Kapton beam window, the 10 μm Kapton detector window and cathode, the 35 μm
copper anode and readout strips, the 125 μm PCB and the gas filled spaces of the
detector.

Table 6.1 summarizes the calculated energy loss for different detector inclinations
with respect to the beam. The ratio between the energy loss in the first and in the

Table 6.1 Mean energy loss in keV of protons with an initial energy of 20 MeV in the 6.0 mm wide
drift gaps of the two Micromegas layers

	0°	10°	20°	30°	40°
Micom 0	17.9 ± 0.4	18.7 ± 0.4	19.5 ± 0.4	21.1 ± 0.4	24.0 ± 0.4
Micom 1	20.7 ± 0.4	21.2 ± 0.4	22.5 ± 0.4	24.8 ± 0.4	28.8 ± 0.4

Calculated with a SRIM based TRIM calculation for an Ar:CO_2 93:7 vol% gas mixture at 20 °C
and 1013 mbar Ziegler et al. (2010)

second detector layer is approximately equal for all angles. The absolute energy loss increases approximately like $1/\cos\vartheta$ for increasing inclination ϑ.

By measuring the current between mesh and anode strips at high rate and using $(52 \pm 2)\%$ electron transmission through the mesh (cf. Sect. 2.6), a gas gain of (1330 ± 300) has been calculated for the first detector layer at $E_{amp} = (33.3 \pm 0.2)\,\text{kV/cm}$ and $E_{drift} = (0.33 \pm 0.02)\,\text{kV/cm}$. For the second layer, a gain of (913 ± 230) is measured at $E_{amp} = (31.7 \pm 0.2)\,\text{kV/cm}$.

The gas gain for a specific detector gas is a function of the amplification gap width, the electric amplification field and the gas pressure. By comparing the ratio of measured gas gains for the two detector layers with the ratio of expected gain, given by Eq. (2.21), the amplification gap width of the second detector layer can be determined. Note, that the amplification gap width d enters twice into that relation: via the amplification field $E_{amp} = U_{amp}/d$, which influences the value of the Townsend coefficient $\alpha(U_{amp}, d, p)$ and directly in the exponent. The amplification gap width of the first layer is defined by the thickness of the solder resist layers, that constitute the mesh supporting structure. Starting from the assumed gap width $d = (150 \pm 1)\,\mu\text{m}$ of the first layer, a width of $d = (164 \pm 1)\,\mu\text{m}$ for the second layer is determined. The larger amplification gap is due to a slight swelling of the mesh supporting pillars, that is also visible under the microscope and has been caused by a long etching process with old solvent.

It should be noted, that the absolute gas gain values, quoted above, are only approximate, large error contributions come from the limited current monitoring accuracy of the high voltage supply, the disregard of long-time charge-up and the uncertainty of the particle rate determination.

The pulse height is shown in Fig. 6.3 as a function of the amplification field. It was measured with low-rate and perpendicularly incident 20 MeV protons and is directly

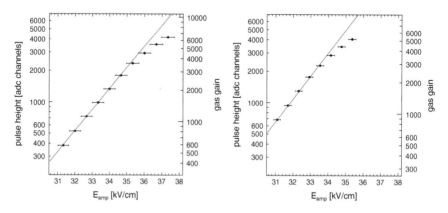

Fig. 6.3 Pulse height as a function of the amplification field E_{amp} for the first (*left*) and the second detector layer (*right*), measured at a drift field of $E_{drift} = 0.33\,\text{kV/cm}$ with perpendicularly incident protons at a mean rate of $(420 \pm 10)\,\text{Hz}$. The gas gain is given by the second axis, it should be noted that its absolute scale is afflicted with an uncertainty of 25 %. The gain in the second layer is higher due to a larger amplification gap, the pulse height is again higher due to a larger proton energy loss

Fig. 6.4 Pulse height in the first detector layer as a function of the drift field for different proton rates, measured at an amplification field of $E_{amp} = 31.3$ kV/cm with perpendicularly incident protons. Assuming homogeneous particle distribution within the beam spot, the rates correspond to fluxes of (1.4 ± 0.3) kHz/cm^2, (1.3 ± 0.2) and (4.0 ± 0.6) MHz/cm^2

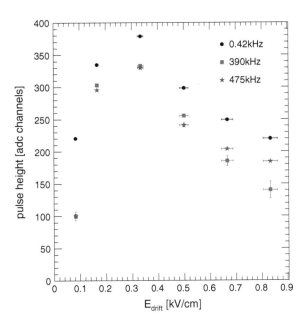

proportional to the gas gain, which is shown by the second axis. An exponential has been drawn to guide the eye. For increasing amplification field, the measured points are slightly lower than the expected exponential behavior. This expected behavior is caused by the limited dynamic range of the APV25 front-end chip. Due to the larger amplification gap of the second layer, its gas gain is for equal amplification fields by a factor of 1.76 higher than in the first layer. The pulse height is by a factor 2.1 higher, due to the larger gain and additionally the larger proton energy loss in this layer.

The ~8 mbar overpressure, used to press the readout structure against the mesh, leads to a slight deformation of the readout structure on the order of 2 μm between the mesh supporting pillars. Over the detector, spatial variations of the pulse height on the order of 15 % are observed.

In Fig. 6.4 the pulse height for different particle rates is shown as a function of the drift field. The physical relevant quantity is not the particle rate but the particle flux density. As discussed in Sect. 6.1.2, the absolute flux density, especially for high rates, is difficult to determine, due to the large flux density inhomogeneity caused by the beam absorbers. In order to avoid large systematic uncertainties in the discussion of the results, the rates are used for data description. The flux densities, quoted in the caption of Fig. 6.4 are lower bounds for the true flux densities.

A pulse height decrease of 15 % is observed when increasing the rate from 0.42 to 475 kHz.

In Fig. 6.5 the pulse height as a function of the drift field is shown for perpendicularly incident 20 MeV protons. The typical behavior for Micromegas is observed, see Sect. 2.6: With increasing drift field the pulse height increases to a maximum

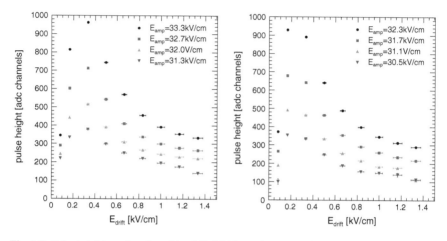

Fig. 6.5 Pulse height as a function of the drift field for various amplification fields for the first (*left*) i.e. the upstream, and the second (*right*) detector layer, measured with perpendicularly incident 20 MeV protons at a mean rate of (420 ± 10) Hz

at $E_{drift} = 0.3$ kV/cm, that corresponds to the minimum of the transverse electron diffusion, see Fig. 2.5. For further increasing drift field, the pulse height decreases again. This is due to the decreasing mesh electron transparency and the increasing transverse diffusion. Both effects lead to a loss of ionization charge on the mesh.

The ratio $ph_{max}/ph(1.3 \text{ kV/cm}) \sim 3.1 \pm 0.3$ is considerably larger than the ratio $\sim 1.7 \pm 0.2$ observed with the same gas mixture in high-energy pion and muon beams. This can be explained by the thicker wires of the woven stainless steel mesh, that has been used in the construction of this small floating strip Micromegas doublet. The increasing transverse diffusion leads to a larger charge loss as the overall mesh thickness increases. This is in agreement with a GARFIELD based mesh transparency simulation, Sect. 2.6.

In Fig. 6.6 the pulse height for different track angles of 10°, 20°, 30° and 40° is shown. As the proton path length increases with increasing track inclination $\propto 1/\cos \vartheta$, the ionization and thus the pulse height increases with the angle. The ionization yield, calculated from the energy loss, given in Table 6.1, has been superimposed. It has been scaled using the relation between the pulse height, the gas gain and the corresponding ionization charge, that is displayed in Fig. 6.3. To account for the pulse height decrease due to the higher particle rates in the angular measurements, the ionization yield has been furthermore scaled with a factor (0.85 ± 0.05). This corresponds to the observed pulse height decrease with increasing rate, cf. Fig. 6.4.

The underlying trend of the calculated ionization yield is reproduced by the measured pulse heights, although considerable deviations are visible. These deviations can partly be explained by the fluctuating particle rates, that may lead to non-proportional variation of the particle fluxes, as discussed in Sect. 6.1.2. It should be noted, that the deviations from the expectation are in the second layer smaller than in the first layer. This is due to a spatial homogenization of the particle flux

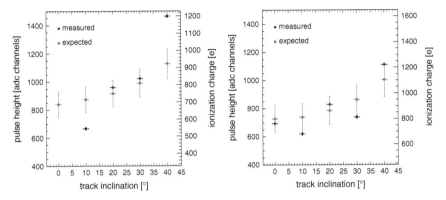

Fig. 6.6 Pulse height as a function of the track inclinations for the first (*left*) and the second detector layer (*right*). Measured with 20 MeV protons with amplification fields of $E_{amp0} = 33.3$ kV/cm and $E_{amp1} = 31.7$ kV/cm and drift fields $E_{drift0,1} = 0.18$ kV/cm. The particle rates were (445 ± 10) kHz and (460 ± 10) kHz for the $10°$, (360 ± 30) kHz and (420 ± 10) kHz for the $20°$, (520 ± 15) kHz and (510 ± 15) kHz for the $30°$ and (470 ± 10) kHz and (490 ± 10) kHz for the $40°$ measurements

in the second layer, caused by multiple scattering of the beam particles. Additional deviations are due to spatial variations of the pulse height on the order of 15 %.

The pulse height for different track angles as a function of the drift field is shown in Fig. 6.7. For track inclinations above $20°$, the most probable pulse height begins to vanish in the electronic noise for $E_{drift} > 0.5$ kV/cm. This is due to the increased

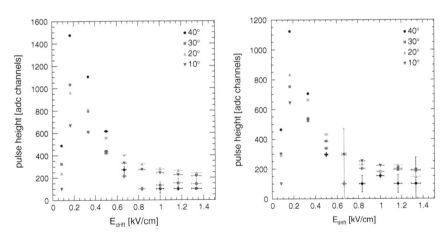

Fig. 6.7 Pulse height as a function of the drift field for different track inclinations for the first (*left*) and the second (*right*) detector layer. Measured with 20 MeV protons with amplification fields of $E_{amp0} = 33.3$ kV/cm and $E_{amp1} = 31.7$ kV/cm. The mean particle rates were (470 ± 20) kHz for the $10°$, (435 ± 15) kHz for the $20°$, (530 ± 15) kHz for the $30°$ and (490 ± 20) kHz for the $40°$ measurements

mesh opacity with increasing drift field and the larger spread of ionization charge over more strips with increasing angle.

6.1.4 Efficiency

In the following section, the efficiency in both detector layers as a function of the drift and amplification field and for different track inclinations will be discussed. It is determined with the method, described in Sect. 4.6: Events with a hit in detector layer i are selected and counted and the number of events, in which *also* layer j registered a hit is counted. The efficiency is given by the ratio of the two numbers.

Statistical errors are calculated according to Eq. (4.18). If error bars are not visible in the figures in the following section, the markers are larger than the errors.

The hit efficiency of both detector layers is shown in Fig. 6.8 as a function of the drift field for different amplification fields. In the drift field region $0.15\,\text{kV/cm} \leq E_{\text{drift}} \leq 0.6\,\text{kV/cm}$ efficiencies above 0.96 are reached for all amplification fields. In this region the efficiency depends weakly on the amplification field. For the highest amplification fields, maximum efficiencies above (0.997 ± 0.001) and (0.987 ± 0.001) for layer 0 and layer 1 are reached, respectively.

The optimum efficiencies of the second layer are by 0.01 smaller than in the first layer. This is due to 3 inefficient strips in the region irradiated by the beam halo. For $E_{\text{drift}} > 0.6\,\text{kV/cm}$ the efficiencies decrease with increasing drift field as the decreasing mesh electron transparency and the increasing transverse electron diffusion lead to a loss of ionization charge at the mesh. In the first detector layer, the decrease for the lower three amplification fields is stronger than in the second.

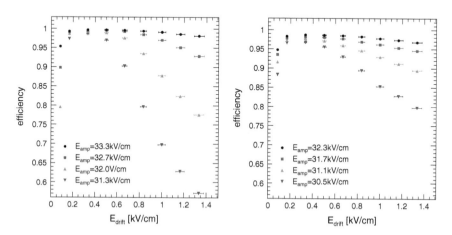

Fig. 6.8 Hit efficiency of the first i.e. the upstream detector (*left*) and the second detector layer (*right*) as a function of the drift field for different amplification fields. Measured with perpendicularly incident 20 MeV protons at a rate of (420 ± 10) Hz

Fig. 6.9 Reconstructed hit positions for both detector layers at drift fields of $E_{drift} = 1.33\,kV/cm$ and amplification fields of $E_{amp0} = 31.3\,kV/cm$ and $E_{amp1} = 30.5\,kV/cm$. Taken from the runs that correspond to the blue triangles at highest drift field in Fig. 6.8

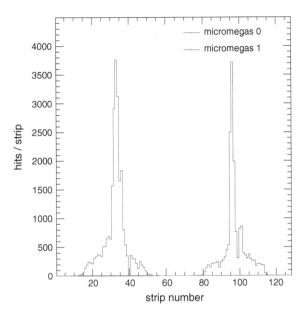

This is a deficiency of the signal identification algorithm and can be understood as follows:

Due to the thin beam spot, with only (1.3 ± 2) mm height, three strips are dominantly hit in the first layer. Multiple scattering of the low energy protons broadens the beam considerably, leading to six dominantly hit strips in the second layer, see Fig. 6.9.

The algorithm, used to calculate the signal standard deviation on each strip, relies on a sufficiently large number of alternately hit strips. Otherwise signals contaminate the standard deviation calculation and too large values are determined. Since hit strips are identified by selecting signals $>3\sigma$, small signals are thus incorrectly missed. As can be seen in Fig. 6.10, the standard deviations of the dominantly hit strips are considerably larger than the correct value around 13 ADC channels. This effect is stronger for the first detector layer, which leads in combination with the thinner beam spot to a stronger efficiency drop for high drift fields, where typically only one to two strips are hit per event.

In Fig. 6.11 the hit efficiency as a function of the drift field is shown for the different detector inclinations of $10°$, $20°$, $30°$ and $40°$. For all angles, efficiencies above 0.97 and up to 0.995 are reached for $0.15\,kV/cm \leq E_{drift} \leq 0.4\,kV/cm$. For perpendicular incidence, the efficiencies for the applied amplification fields are well above 0.97 for the first and above 0.95 for almost all applied drift fields, see Fig. 6.8. The observed efficiency decrease for increasing angles for higher drift fields is caused by the larger spread of the ionization charge over more strips.

Unexpectedly, the decrease is strongest in both detector layers for the $30°$ track inclination measurements. This can be correlated to an upward fluctuation of the

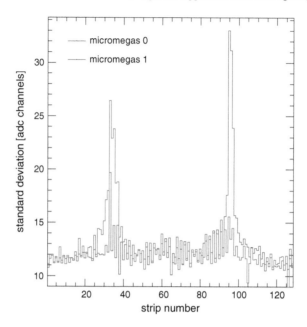

Fig. 6.10 Signal standard deviations for all strips in the two detector layers. Same runs as in Fig. 6.9. Clearly visible are the elevated standard deviations of the dominantly hit strips. The alternating substructure is caused by different stripline lengths in the detectors and on the APV25 front-end board

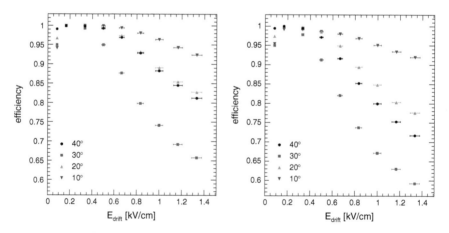

Fig. 6.11 Hit efficiency of the first i.e. the upstream detector (*left*) and the second detector layer (*right*) as a function of the drift field for different track inclination angles. Measured with 20 MeV protons with amplification fields of $E_{amp0} = 33.3\,\text{kV/cm}$ and $E_{amp1} = 31.7\,\text{kV/cm}$. The mean particle rates were $(470 \pm 20)\,\text{kHz}$ for the $10°$, $(435 \pm 15)\,\text{kHz}$ for the $20°$, $(530 \pm 15)\,\text{kHz}$ for the $30°$ and $(490 \pm 20)\,\text{kHz}$ for the $40°$ measurements

particle rate, which can lead to a strong increase of the particle flux, as discussed in Sects. 6.1.2 and 6.1.3. This decreases the absolute pulse height, which again leads to a lower efficiency. Note, that at high drift field, the most probable pulse height is small (cf. Fig. 6.7) and close to the noise threshold, such that a small pulse height decrease on the order of 10 % can lead to strong variation of the efficiency.

6.1.5 Angular Resolution

In the following section the single plane angular resolution of the two Micromegas layers is discussed. The μTPC-reconstruction method, introduced in Sect. 4.8, is used. The ionization charge arrival time is extracted by fitting the signal rising edge with an inverse Fermi function, see Sect. 4.3. The electron drift velocities were calculated with MAGBOLTZ (Biagi 1999), see Fig. 2.4. It should be noted, that all events in which both detector layers registered a hit, are used in the presented analysis. The only constraint on signal selection was the demand of at least two hit strips per layer, leading to a reconstruction efficiency of 99.8 % at the working point with respect to the simple hit constraint.

In Fig. 6.12 the most probable reconstructed angle and the angular resolution is shown as a function of the incident beam angle. The most probable reconstructed angle is extracted by fitting the central peak in the inclination spectrum with a symmetric Gaussian function. Occasional misinterpretation of noise as signal and capacitive coupling of signals on neighboring strips lead to reconstructed angles, that are

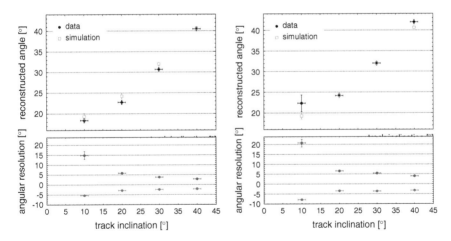

Fig. 6.12 Reconstructed track inclination and angular resolution of the first detector layer (*left*) and the second detector layer (*right*) as a function of the true track inclination. Measured with amplification fields $E_{amp0} = 33.3$ kV/cm and $E_{amp1} = 31.7$ kV/cm and a drift field of $E_{drift0,1} = 0.17$ kV/cm. The angular accuracy for angles higher than the most probable value is represented by positive values and vice versa

higher than the most probable value, see Sect. 4.8.4 for a discussion. The angular resolutions for track angles lower and higher than the most probable value are thus treated separately. Both values are extracted by fitting the peak with an asymmetric Gaussian function, Sect. 4.8.3.

The behavior of the measured most probable reconstructed angle can in both detectors be well explained by the capacitive coupling of signals on neighboring strips (see Sect. 4.8.4), predicting a deviation of $+(9.3 \pm 0.8)°$, $+(4.3 \pm 0.7)°$, $+(2.0 \pm 0.5)°$ and $+(0.6 \pm 0.2)°$ for the four investigated angles. The resolution improves with increasing track inclination as more strips are hit i.e. the lever arm increases and negative effects due to the finite time resolution and capacitive coupling of adjacent strips decrease.

The measured angular resolution in the first detector layer is for all track inclination better than in the second layer. For low track inclination, this is dominantly caused by the unintentionally used higher gas gain and pulse height in the first detector layer.

A Monte Carlo simulation shows, that the beam emittance increases due to multiple scattering of the low energetic protons in the two readout structures by $(1.6 \pm 0.2)°$ for $10°$ track inclination up to $(1.8 \pm 0.2)°$ for $40°$ track inclination. This fakes a worse angular resolution in the second layer and explains well the observed difference in resolution between the two layers at larger track inclination.

An additional possibility to characterize the quality of the reconstruction is to determine the fraction of reconstructed track angles inside and outside the one- and two-sigma band around the most probable reconstructed angle. The inclination distribution clearly has non-Gaussian tails, more pronounced towards larger reconstructed angles, except for a track inclination of $10°$, where the tails extend rather to smaller and negative angles. A detailed compilation of the fraction within certain intervals can be found in the Appendix C.1.1.

The most probable reconstructed angle and the angular resolution as a function of the amplification field and thus the gas gain can be seen in Fig. 6.13.

The most probable reconstructed angle decreases and approaches the true track inclination for increasing gas gain up to an amplification field of $E_{amp} \sim 32.5\,kV/cm$ for the first and $E_{amp} \sim 31.0\,kV/cm$ for the second layer. This is due to a better signal discrimination against noise, but is to a smaller extent also an artifact of the most probable value determination, influenced by the asymmetry of the angle distribution.

Between the two detector layers, an offset of about $1°$ is observed. Since only measurements with positive track inclination were performed, it is not clear, whether this is caused by an angle between the two detector layers, or whether it is an artifact of the most probable value determination, distorted by the more asymmetric central peak in the track inclination spectrum of the second layer. A Monte Carlo simulation of multiple scattering in the readout structures of the doublet detector, showed, that the mean track inclination is not distorted by multiple scattering.

The angular resolution improves with increasing gas gain, enabled by a higher single strip efficiency and better temporal resolution of the higher signals.

In Fig. 6.14 the most probable reconstructed angle and the angular resolution are shown as a function of the drift field in the first detector layer for the four track

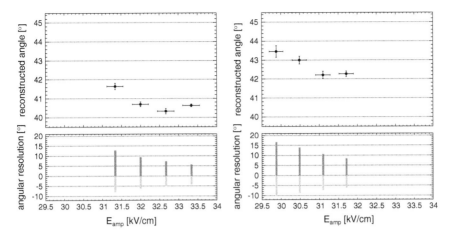

Fig. 6.13 Reconstructed track inclination and angular resolution of the first detector layer (*left*) and the second detector layer (*right*) as a function of the amplification field. The angular accuracy for angles higher than the most probable value are represented by positive values and vice versa. Measured at a track inclination of $(40 \pm 1)°$ with $E_{drift} = 0.33$ kV/cm

inclinations $(10 \pm 1)°$, $(20 \pm 1)°$, $(30 \pm 1)°$ and $(40 \pm 1)°$. As expected and discussed in Sect. 4.8.4, the most probable reconstructed angle is always larger than the true track inclination. For the optimum drift field $E_{drift} = 0.17$ kV/cm, where also the angular resolution reaches its best value, this can be fully explained by the capacitive coupling of neighboring strips. See also Fig. 6.12 for reference.

The increase of the most probable angle and also of the resolution for drift fields lower and higher than the optimum value is caused by two effects: First, noise misinterpreted as signal on non-hit neighboring strips always leads to an shift of the reconstructed angles to larger values. This is correlated to the decreasing pulse height due to recombination and lower mesh transparency, as discussed in Sect. 6.1.3. Second, the electron drift velocity increases for drift fields $E_{drift} > 0.17$ kV/cm (Fig. 2.4 and Sect. 6.1.7), resulting in a larger relative inaccuracy of the signal time resolution. It should be noted, that mainly due to the first effect, the angular reconstruction is very inaccurate for $10°$ track inclinations for $E_{drift} \gtrsim 0.5$ kV/cm.

The intention of the presented measurements was to study the underlying effects, being responsible for the observed behavior of the most probable reconstructed angle and the angular resolution.

6.1.6 Spatial Resolution

Since only two tracking detectors were used in the measurements, the spatial resolution can be estimated by assuming parallel tracks for all beam particles and measuring

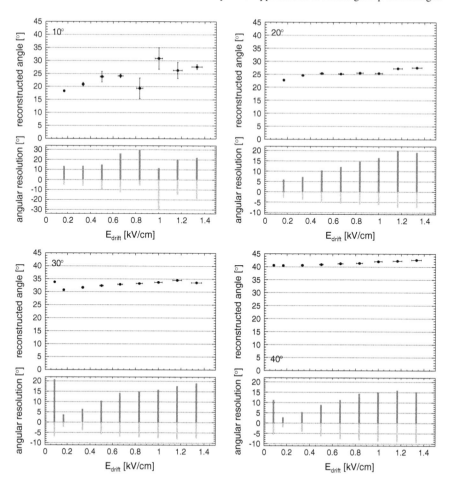

Fig. 6.14 Reconstructed track inclination and angular resolution of the first detector layer as a function of the drift field for true track angles of $(10 \pm 1)°$ (*top left*), $(20 \pm 1)°$ (*top right*), $(30 \pm 1)°$ (*lower left*) and $(40 \pm 1)°$ (*lower right*). The angular accuracy for angles higher than the most probable value are represented by positive values and vice versa. Measured with 20 MeV protons at mean particle rates of (470 ± 20) kHz for the 10°, (435 ± 15) kHz for the 20°, (530 ± 15) kHz for the 30° and (490 ± 20) kHz for the 40° measurements. Note the different scale for the angular resolution for 10° track inclination. The behavior in the second layer is equivalent

the difference of the reconstructed hit positions in the first and the second detector layer

$$\Delta x = x_0 - x_1 . \tag{6.1}$$

Calculating this residual for many similar tracks yields a nearly Gaussian residual distribution, whose width σ_{res} can be determined from a fit with a Gaussian function. The spatial resolution $\sigma_{SR,0}$ and $\sigma_{SR,1}$ of the first and the second layer, respectively are then given by

$$\sigma_{res} = \sqrt{\sigma_{SR,0}^2 + \sigma_{SR,1}^2} \, , \tag{6.2}$$

simplifying to

$$\sigma_{SR} = \frac{\sigma_{res}}{\sqrt{2}} \, , \tag{6.3}$$

assuming equal spatial resolution in both layers.

Note, that the necessary initial assumption of parallel tracks does definitely not hold for these measurements, due to the beam divergence and the significant multiple scattering of the low energy 20 MeV protons in the beam pipe window, air, and the detectors themselves. The determined values are thus largely overestimated, nevertheless several trends are observed.

For measurements with low proton rate of 420 Hz, the determined spatial resolution of the two layers is on the order of (520 ± 20) µm. Increasing the particle rate from 0.42 to 390 kHz, leads to a deterioration of the spatial resolution by 10 %. A beam spot, elongated into the direction of the strips, as in these two measurements, can lead to an additional degradation of the measured spatial resolution: Since no hit position information in strip direction is possible, a mutual rotation of the two detector planes cannot be corrected for. Before the high-rate measurements and the study with inclined detectors, the beam has thus been refocused to a more circular shape. An improvement of the spatial resolution to (475 ± 20) µm when moving to highest particle rate and elliptic beam focus reveals, that indeed a mutual rotation of readout planes is present.

A detailed discussion about the dependence of the observed spatial resolution of the drift and amplification fields, the particle rate and the track inclination, can be found in the Appendix C.1.2.

6.1.7 Measurement of Electron Drift Velocities

By inverting the µTPC reconstruction method, the electron drift velocity can be calculated from the known track inclination. The method has been introduced in Sect. 4.8.5. The systematic deviations in the µTPC reconstruction method are equally present in this method, i.e. due to capacitive coupling of neighboring anode strips, the reconstructed µTPC line slopes are too small resulting in too large values for the reconstructed drift velocity as a function of the true track inclination.

In Fig. 6.15 the reconstructed electron drift velocities as a function of the drift field are shown for both detector layers and all four track inclination angles.

For a track inclination of $10°$, the method does not work properly, as it has been discussed in Sect. 6.1.5. This leads to significantly overestimated values for the drift velocity.

As discussed in Sect. 4.8.5, the actual drift gap width can be determined by comparing measured and expected electron drift velocities. For the presented measurements, no deviation of the drift gap width from the expected value of 6 mm is visible,

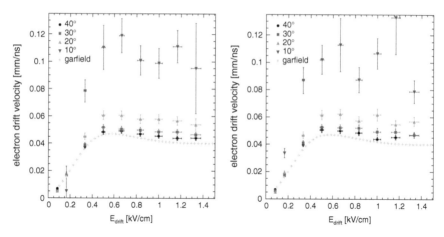

Fig. 6.15 Reconstructed electron drift velocities of the first (*left*) and the second detector layer (*right*) as a function of the drift field for different track inclination angles. Measured with 20 MeV protons with amplification fields of $E_{amp0} = 33.3\,\text{kV/cm}$ and $E_{amp1} = 31.7\,\text{kV/cm}$

although the detector is slightly pressurized with respect to the surrounding. This has been achieved in the construction by separating the drift cathode and the gas window.

6.1.8 Discharge Behavior

The formation of discharges between anode strips and micro-mesh has been discussed in Sect. 2.7. The critical charge density of approximately 2×10^6 e/0.01 mm^2 can be exceeded during high-flux proton irradiation due to events with large energy transfers to the detector gas. Since each measurement run with a certain parameter set lasted only for two minutes and the discharge probability is rather small, the discharge probability could not be determined reliably for the whole electric field parameter space.

Approximate discharge probabilities on the order of $(3 \pm 1) \times 10^{-7}$ per incident particle have been determined. The observed dependence of the discharge probability on the drift field and the track inclination can be found in the Appendix C.1.3.

Cooperative effects due to several particles crossing the same region in rapid succession can not be excluded. In later measurements with protons and carbon ions at rates of up to 2 GHz, no cooperative effect are observed though, see Sect. 6.2.

6.1.9 Summary

A floating strip Micromegas doublet with low material budget has been tested with 20 MeV protons at the tandem accelerator in Garching. Particle tracking with perpendicular and inclined beams was possible for all investigated rates between several 100 Hz and 550 kHz. The mean particle rate has been measured with a scintillator detector every 10 s, the tandem accelerator provided a stable proton beam with rate fluctuations below 20 %.

The dependence of the proton signal pulse height on amplification and drift field has been determined, the expected behavior is observed. A pulse height decrease by 15 % is observed when increasing the particle rate from 0.42 to 475 kHz. With increasing detector inclination and correspondingly increasing proton track length, the pulse height increases as expected.

For optimized electric fields, the detection efficiency in both detectors is above 0.98 for all rates and all detector inclinations.

The track inclination determination in a single detector plane has been investigated with both Micromegas. For optimized field parameters the most probable reconstructed angles agree well with the expected values. The description of the monotonous systematic deviations presented in Sect. 4.8.4 is shown to be valid. A correction of the most probable reconstruction angle is possible. The angular resolution has been investigated as a function of the track inclination. An optimum resolution of $\left(^{+3}_{-2}\right)^{\circ}$ is reached for a track inclination of 40°.

The determination of the spatial resolution is strongly limited by the beam divergence and multiple scattering of protons in the detectors, as only two detectors were used in the test measurements. Optimum resolutions of 480 μm are reached for the highest rate at perpendicular beam incidence. For beams with inclinations above 18°, the alternative hit reconstruction with the μTPC method yields by 15–20 % more accurate results than the usual charge-weighted mean reconstruction.

Electron drift velocities have been measured by inverting the μTPC reconstruction. The reconstructed values are subject to the expected systematic deviations.

Approximate discharge probabilities on the order of 3×10^{-7} have been determined, no influence on the detector performance is observed, as expected.

6.2 Floating Strip Micromegas Tracking System for High-Rate Proton and Carbon Beam Diagnostics

In the following, a 2 day measurement with a Micromegas tracking system in high-rate proton and carbon ion beams at the Heidelberg Ion Therapy center (HIT) is presented. The detectors are intended for particle tracking in ion transmission imaging (see Sect. 1.3.2). The presented ion beam characterization measurements allow, furthermore, for an investigation of their performance in high-rate ion beams.

The tracking system consisted of a thin floating strip Micromegas doublet and a resistive strip Micromegas. Tracks of high rate protons and carbon ions were used to study the detector performance and optimize track reconstruction algorithms. Measurements with fixed beam position were taken using protons with energies between 48 and 221 MeV/u and particle rates of 80 MHz–9 GHz, and with carbon ions with energies between 88 and 430 MeV/u at rates of 2–80 MHz. Moreover, with both ion types, several measurements using the active raster scanning beam delivery system of HIT were performed (Haberer et al. 1993) i.e. the beam position was moved in small steps to different raster points. In medical ion therapy irradiation, the irradiated volume is subdivided into raster points. They are defined by a specific combination of beam position, beam focus, particle energy and flux.

The good multi-hit resolution of Micromegas allows for a characterization of the bunch structure and the spill structure of the particle beam. The track reconstruction method and multi-hit capabilities of Micromegas are presented. The dependence of the pulse height of charge signals on drift and amplification field and on particle rates and energies is discussed. The single particle detection efficiency and the overall detector availability is presented. The achieved spatial resolution in carbon beams is shown. Energy and beam dependent discharge behavior is discussed.

6.2.1 Setup

A schematic view of the detector system and the readout electronics can be seen in Fig. 6.16. Ion hit positions were measured with three Micromegas detectors: A floating strip Micromegas doublet with one-dimensional strip readout and a resistive strip Micromegas with two perpendicular layers of readout strips.

The doublet detector, consisting of two back-to-back floating strip Micromegas with an active area of $6.4 \times 6.4 \, cm^2$, has been described in detail in Sect. 3.4.2. Performance of the detector in 20 MeV proton beams is presented in Sect. 6.1. The readout structure of each layer consists of 128 copper anode strip with 500 μm pitch and 300 μm width. Each anode strip is individually connected via a 22 MΩ SMD resistor to high-voltage. The strips were oriented in horizontal y-direction, such that ion hit positions could be precisely measured in vertical x-direction.

The resistive strip Micromegas with an active area of $9 \times 9 \, cm^2$ and 358 strips per layer provided two-dimensional hit information in x- and y-direction. The resistive strips were oriented in x-direction.

The three gas detectors were read out using APV25 based front-end boards, interfaced by the Scalable Readout System, Sect. 4.1.2. Charge signals, sampled every 25 ns, were acquired for all strips within a readout window of 450 ns around the trigger.

The trigger signal was provided on the first measurement day by a single scintillator, read out with a Hamamatsu R4124 photomultiplier, that was mounted behind the resistive strip Micromegas under an angle of about 20° to the beam axis. Due to the high particle rates of 2 MHz and above, the scintillator detector could not be

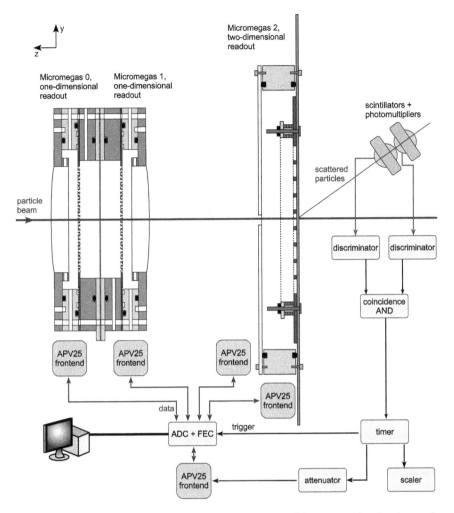

Fig. 6.16 Schematic setup in the high-rate test measurements with protons and carbon ions at the Heidelberg Ion Therapy center, as seen from above. A Micromegas doublet with $6.4 \times 6.4\,\mathrm{cm}^2$ active area and 128 readout strips per layer and a resistive strip Micromegas with two perpendicular strip planes are read out using APV25 front-end boards, interfaced by the Scalable Readout System. Due to the high particle rate, we triggered on scattered particles, creating coincident hits in two scintillators. The attenuated trigger signals itself is acquired with an additional APV25 front-end board to enable correction of the 25 ns time jitter present in the Micromegas signals. Ions are moving into negative z-direction, their hit position in the x-direction is measured by all three micromegas, the resistive strip Micromegas provided hit information in the y-direction

operated in the primary beam. The trigger signal was derived from photons and scattered protons, produced by interaction of the primary beam with the readout structure of the resistive strip Micromegas. In order to eliminate triggers coming from random noise and photons, that are not coincidently produced by ions, traversing the detector system, a second scintillator detector has been added for the second measurement day, thus exclusive triggering on charged secondary particles was possible.

Due to the lack of a high-rate capable trigger detector, the instantaneous particle rate could not be measured directly, such that we rely on the requested beam intensity, as given by the accelerator control system, for rate determination.

Discharge signals in both floating strip Micromegas layers as well as scintillator trigger signals were counted with a FPGA based NIM scaler, read out via RS232 every 10 s. High-voltage for the gas detectors was provided by two 12 channels CAEN A1821 modules (CAEN S.p.A 2013a) in a CAEN SY5527 mainframe (CAEN S.p.A 2013b).

A premixed Ar:CO_2 93:7 vol% gas mixture was flushed through the gas detectors with an overall flux of 2 ln/h. The detectors were operated at atmospheric pressure, which was on the order of 1000 mbar during the measurements.

6.2.2 Accelerator and Beam Characterization

The accelerator complex at the HIT has been described in Sect. 1.3.1. In the following section measurements are discussed, that allow for a characterization of the synchrotron and the microscopic beam structure.

At HIT the active scanning beam delivery system is used for localized tumor irradiation, Sect. 1.3.1. For each iso-energy-slice, which corresponds to a certain depth in the tumor, the high-energy ion beam is steered with fast scanning magnets to a set of raster points, the penetration depth is actively varied by adapting the synchrotron energy. The beam position is measured with multi-wire proportional chambers directly in front of the patient, providing feed-back to the scanning system.

Due to the finite ramping time of the scanning magnets and the necessary integration time of the wire chambers, slow extraction of the beam from the synchrotron is used, Sect. 1.3.1. A particle beam with approximately constant, actively controlled intensity is available over a period of 5 s (Ondreka and Weinrich 2008). The complete cycle length, including injection, ramp up, extraction, chimney and ramp down is energy and particle dependent (Rinaldi 2014) and is on the order of 8 s. The energy dependence is caused by a magnetic field dependent waiting time after acceleration, necessary to cope with Eddy-currents in the magnets. Furthermore, the beam dump phase is longer for higher energy beams. A break with variable duration between two accelerator cycles is used for internal communication between accelerator and beamline subsystems (Schömers 2014). The actual magnetic fields and their temporal behavior in the six main dipole and in the quadrupole magnets of the synchrotron is measured with Hall probes and pickup coils. This allows for a direct correction of the magnet currents, leading to a reduction of the cycle length from the nominal 8.7 s (Feldmeier et al. 2012). A further reduction to about 7 s is foreseen, as the quadrupole magnet hysteresis can in the same way be measured and corrected for.

The spill length and the duration of the breaks between two spills could be measured with the APV25 based data acquisition system: The time, at which each event

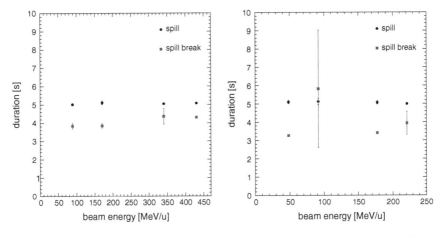

Fig. 6.17 Measured mean duration of beam spill and break between two consecutive spills as a function of the particle energy for carbon ions (*left*) and protons (*right*). Measured with carbon beams with $I_{^{12}C} = 5 \times 10^6$ Hz and proton beams at $I_p = 8 \times 10^7$ Hz. The *error bars* represent the standard deviation of the measured durations for individual spills and spill-breaks per particle energy

is written to disk is recorded in the event header with an accuracy of 1 s. Using the observed number of triggered events per second and comparing it to the mean number of events per second during a spill, the spill and spill-break duration could be determined with sub-second accuracy.

In Fig. 6.17 the spill and spill-break duration is shown for carbon and proton beams as a function of the particle energy. The measurements were performed in continuous beam mode, i.e. a new synchrotron cycle is requested directly after the previous has finished.

The measured spill durations are well compatible with the expected length of 5 s, the spill-breaks increase with particle energy, as expected. Additionally, we observe a considerable variation of the spill-break duration, due to the variable delay between consecutive synchrotron cycles.

The microscopic bunch structure of the beam can be measured in the Micromegas detectors. As discussed in Sect. 6.2.1, the data acquisition system is triggered by a signal in scintillator detectors, created by secondary photons and particles, produced in the readout structure of the resistive strip Micromegas. The Micromegas record particle hits within the acquisition window of 450 ns around the trigger signal. If the bunch spacing i.e. the time between two consecutive bunches is sufficiently small, ions in the previous and in the following bunch can be detected, additionally to the particles in the bunch incorporating the triggering particle. This is possible due to the good multi-hit capabilities of Micromegas detectors.

The signal timing of all strips, corresponding to a cluster, that could be matched to a particle track, is shown in Fig. 6.18. The timing information has been determined from a fit of each strip signal with a skew Gaussian function, Sect. 4.3.

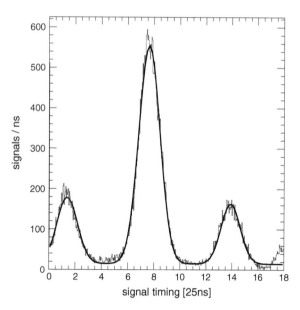

Fig. 6.18 Signal timing distribution, measured in Micromegas 0 with carbon ions at $E_{^{12}C} = 340.77\,\text{MeV/u}$ and $I_{^{12}C} = 5 \times 10^6\,\text{Hz}$

The dominant peak in the spectrum corresponds to particles from the bunch, that was triggered on, particles from the previous and two following bunches are additionally visible. For a fixed number of particles per bunch and an unbiased event selection, the peaks should be similarly high. The inequality, that is observed for all beam intensities, is caused by two effects: First, we predominantly trigger on bunches containing more particles, since the trigger is derived from detection of secondary photons and particles. Second, bunches seem not to be filled homogeneously, which can be plausibly explained by the RF-Knockout extraction method, used at the HIT synchrotron, Sect. 1.3.1.

The spectrum is fitted with a sum of three Gaussian functions and a constant, the distance of the peaks is a fit parameter and allows for an extraction of the bunch spacing.

The expected bunch spacing T can be calculated from the particle energy, represented by $\gamma = E/m_0 c^2$

$$T(\gamma) = \frac{d}{c\sqrt{1 - \frac{1}{\gamma^2}}} \,, \tag{6.4}$$

where d represents the spatial distance of consecutive bunches in the synchrotron. By comparing measured and expected bunch spacing, a distance $d = 31.7\,\text{m}$ has been determined. From the total synchrotron circumference of about 65 m (Ondreka and Weinrich 2008), we can conclude, that the synchrotron is filled with two bunches, which is consistent with reality.

The measured and expected bunch spacing is shown in Fig. 6.19 as a function of the energy for carbon and proton beams. The measured values in carbon beams agree

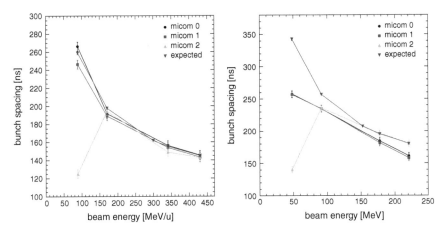

Fig. 6.19 Measured and expected bunch spacing as a function of the particle energy for carbon ions with $I_{12C} = 5 \times 10^6$ Hz (*left*) and protons with $I_p = 8 \times 10^7$ Hz (*right*)

with the expected bunch spacing for energies above 150 MeV/u, corresponding to a bunch spacing below 200 ns. Below 150 MeV/u, deviations are observed, that are caused by the limited width of the data acquisition window. The stronger deviations in the resistive strip Micromegas are caused by a shift of the measured signal timings. The previous bunch is not visible in the distribution anymore, leading to a less stable determination of the bunch spacing. The shift is due to the faster signal rise time in this detector. As the signal rise time is determined by the drift of positive ions from the anode towards the mesh (Sect. 2.5) its smaller amplification gap width of 128 μm compared to 150 and 164 μm in the floating strip detectors leads to a smaller rise time and thus a shift of the signal maximum towards earlier values.

The same holds for the measured bunch spacing in proton beams, where the overall deviation for the floating strip detectors is larger, due to the significantly higher number of particles per bunch. Due to the higher granularity of the resistive strip readout structure, its deviation is smaller for the higher beam energies.

Using the bunch spacing, given by Eq. (6.4), and the mean beam intensities I as given by the accelerator control system, the mean number of particles per bunch $n_{p/b}$ can be calculated

$$n_{p/b} = IT(\gamma). \tag{6.5}$$

Note, that this value depends both on the beam intensity I and on the beam energy E. Considering the 450 ns acquisition window, the mean number of particles, visible in the detectors per triggered event, $n_{p/e}$ can be approximated

$$n_{p/e} = n_{p/b}n_{b/e}(T(\gamma)), \tag{6.6}$$

where the number of bunches per event $n_{b/e}$ can be estimated from the bunch spacing and the acquisition window, assuming an effective bunch length of 50 ns.

Table 6.2 Expected number of particles per bunch for carbon beams as a function of the beam intensity and particle energy

Intensity (Hz)	Energy (MeV/u)				
	88.83	170.66	299.94	340.77	430.1
2.00×10^6	0.52	0.40	0.32	0.31	0.29
5.00×10^6	1.30	0.99	0.81	0.78	0.72
1.00×10^7	2.59	1.98	1.62	1.55	1.45
2.00×10^7	5.18	3.96	3.23	3.10	2.90
5.00×10^7	12.96	9.89	8.08	7.76	7.25
8.00×10^7	20.73	15.83	12.93	12.42	11.60

Table 6.3 Approximated expected number of particles per event for carbon beams as a function of the beam intensity and particle energy

Intensity (Hz)	Energy (MeV/u)				
	88.83	170.66	299.94	340.77	430.1
2.00×10^6	1.26	1.59	1.59	1.59	1.64
5.00×10^6	1.94	2.48	2.49	2.47	2.59
1.00×10^7	3.89	4.95	4.59	4.50	4.64
2.00×10^7	7.77	9.89	9.18	9.00	9.28
5.00×10^7	19.43	24.73	22.96	22.51	23.20
8.00×10^7	31.09	39.56	36.73	36.01	37.11

The accuracy of the shown values is better than 10%, assuming a homogeneous distribution of particles over all bunches

In Tables 6.2 and 6.3, the mean number of particles per bunch and per event, respectively, are given for carbon beams for the intensity and energy parameters, that were used in this measurement.

For the four smallest intensities used, reliable single particle tracking for almost all particles is expected to be possible. A carbon ion typically produces a signal on three neighboring strips. As long as two particles do not hit the same strip group, they can be separated. For the higher rates, single particle tracking is still possible, but later arriving particles or those directly in the center of the beam are expected to be inseparable.

The mean number of protons per bunch and per event are given in Tables 6.4 and 6.5. It is clear, that, depending on the beam diameter, for the lowest particle rate single particle tracking is still possible, but a considerable fraction of protons will merge into conjoint clusters. For intensities above 10^8 Hz, single particle tracking is becoming increasingly difficult, for intensities above 2×10^8 Hz, single particle tracks cannot be resolved anymore, but the beam position can be measured accurately.

Table 6.4 Expected number of particles per bunch for proton beams as a function of the beam intensity and particle energy

Intensity (Hz)	Energy (MeV/u)				
	48.12	91.48	152.58	177.7	221.06
8.00×10^7	27.4	20.5	16.6	15.6	14.4
2.00×10^8	68.5	51.3	41.5	39.1	36.0
4.00×10^8	137.1	102.6	82.9	78.1	72.0
8.00×10^8	274.1	205.3	165.8	156.2	144.0
2.00×10^9	685.3	513.2	414.5	390.6	360.0

Table 6.5 Approximated expected number of particles per event for proton beams as a function of the beam intensity and particle energy

Intensity (Hz)	Energy (MeV/u)				
	48.12	91.48	152.58	177.7	221.06
8.00×10^7	27.4	34.9	36.5	35.9	35.3
2.00×10^8	68.5	87.2	91.2	89.8	88.2
4.00×10^8	137.1	174.5	182.4	179.7	176.4
8.00×10^8	274.1	349.0	364.8	359.3	352.8
2.00×10^9	685.3	872.4	912.0	898.3	882.0

The accuracy of the shown values is better than 10%, assuming a homogeneous distribution of particles over all bunches

6.2.3 Track Reconstruction and Multi-Hit Resolution

Particle hits in the individual Micromegas layers are reconstructed by merging the position and charge information from neighboring strips. The reconstructed hit positions are then corrected for systematic deviation caused by the charge distribution discretization due to the periodic strip structure, see Sect. 4.3. Since usually more than one valid track is present in the detector system, the Hough transform based track reconstruction algorithm is used to identify the hit positions, belonging to specific particle tracks (Sect. 4.4.2).

The track reconstruction algorithm searches for perpendicularly incident[3] and straight tracks over the entire active area of the detectors. It yields the combination of clusters, belonging to a certain track and an estimate of the track slope and intercept, that is not further used. Note that the term cluster denotes in this context a group of neighboring hit strips.

The identified tracks are then fitted analytically with a straight line to determine the track parameters, see Sect. 4.4.3. In the pulse height-, signal timing- and spatial resolution-analysis, except for a few exceptions, clusters are considered only if they can be matched to track.

---

[3] Track inclination angle between $-3°$ and $1°$.

As discussed in Sect. 6.2.2, particles that coincidently create a signal on the same group of strips cannot be separated in a strip detector. A simplified Monte Carlo simulation has been developed to predict the reconstructible number of particles. It incorporates the detector strip geometry, the cluster width in the detector, the beam spot size and shape and the number of particles per event. The beam shape is described by the sum of two Gaussian functions, corresponding to central beam and beam halo. The width of the central beam is given by $\sigma_{central} = FWHM/2\sqrt{2\ln 2}$, where FWHM is the beam width as given by the accelerator control system, the width of the beam halo is assumed to be $\sigma_{halo} = 2\sigma_{central}$. The distribution of 94 % of all particles is described by the dominant peak. These parameters have been adjusted to reach agreement between measured data and simulation.

In Fig. 6.20 the measured beam spot is shown for two different beam intensities. Due to the high flux density in the center of the beam, signals from single particles merge into conjoint clusters. An increase of the observed beam width is visible, that is caused by particles in the beam halo.

In Fig. 6.21 the mean number of reconstructed hits per event for carbon beams is shown for the first floating strip Micromegas. Superimposed are the simulated mean number of reconstructed clusters and the true number of hits per event, as given by Table 6.3. The mean number of reconstructed hits rises linearly for $I < 20 \times 10^6$ Hz and saturates for higher beam intensities at around 5.5 hits per event. The saturation is due to merging of different particles into conjoint clusters. Due to the biased trigger system, favoring bunches in which the number of particles per bunch is higher than the average, the measured number of hits for lower rates is slightly higher than the simulated one. The number of particles per bunch does not seem to strictly follow a Poisson distribution at lower particle rates, which is plausible due to the

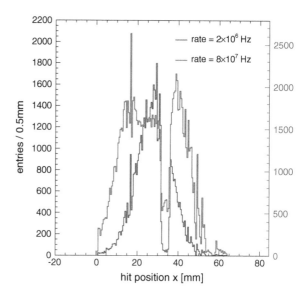

Fig. 6.20 Beam spot in the first layer of the floating strip Micromegas doublet. Measured with 88.83 MeV/u carbon ions at two different beam intensities with $E_{amp} = 30.7$ kV/cm and $E_{drift} = 0.33$ kV/cm. The inefficient region between $x = 31$ mm and $x = 36$ mm is created by a group of 10 dead strips in the second floating strip Micromegas layer. A clear increase of the beam width due to particles in the beam halo is observed

Fig. 6.21 Mean number of reconstructed hits per event in the first floating strip Micromegas as a function of the beam intensity. Shown are the measured values in carbon beams with $E_{12C} = 88.83$ MeV/u, and the simulated number of true and reconstructed hits. The beam FWHM is 13.4 mm

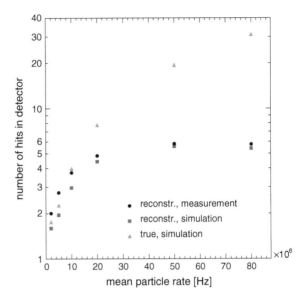

Fig. 6.22 Mean number of reconstructed tracks per event as a function of the beam intensity. The same runs as in Fig. 6.21 are shown. The vertical error bars are smaller than the markers

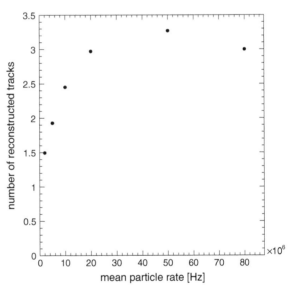

extraction mechanism of the synchrotron. A good agreement between simulation and measurement for the higher intensities is visible, with deviations less than 10 %.

Due to this cluster merging not all reconstructed hits correspond to valid particle hit positions and can be used for tracking. The mean number of reconstructed tracks per event, shown in Fig. 6.22, is thus lower than the mean number of reconstructed hits and even decreases with further increasing intensity for $I > 50 \times 10^6$ Hz.

6.2.4 Pulse Height Behavior

The dependence of the pulse height of signals, produced by traversing protons and carbon ions, on the electric field parameters, on the particle rate and particle energy is discussed in the following section. The dominant peak in the pulse height or energy loss spectrum, which is equivalent to the most probable energy loss of protons and carbon ions, can be fitted with a Gaussian function. The most probable pulse height, as defined by the position of the maximum of the fit function, is used to characterize the pulse height behavior in the following.

The pulse height as a function of the amplification field in the two floating strip Micromegas, measured with 88.83 MeV/u carbon ions, is shown in Fig. 6.23. For constant drift field, the pulse height is directly proportional to the gas gain (Eq. 2.21), such that the observed exponential behavior agrees with the expected one. An exponential function has been drawn to guide the eye. For $E_{amp} > 29$ kV/cm, the measured pulse heights are slightly lower than the expectation. A similar effect has been observed in measurements with 20 MeV protons, Sect. 6.1.3. It is due to saturation effects of the APV25 front-end chip. No significant additional detector saturation effects with increasing gas gain are visible. Since the ionization yield in both detector layers is approximately equal, the higher pulse height in Micromegas 1 is directly correlated to a higher gas gain at equal amplification field, due to a larger amplification gap of (164 ± 1) μm as compared to (150 ± 1) μm for Micromegas 0.

In Fig. 6.24 the pulse height of both floating strip Micromegas is shown as a function of the drift field. Typical behavior is observed: Starting at low field, the

Fig. 6.23 Pulse height of the floating strip detectors as a function of the respective amplification field. Measured with carbon ions with $E_{12C} = 88.83$ MeV/u and $I_{12C} = 2 \times 10^6$ Hz at $E_{drift} = 0.33$ kV/cm

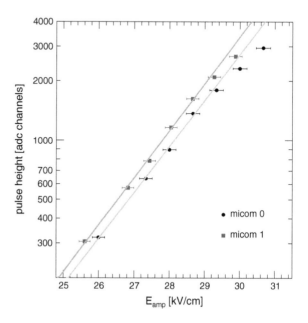

Fig. 6.24 Pulse height of the floating strip detectors as a function of the respective drift field. Measured with carbon ions with $E_{12_C} = 88.83 \, \text{MeV/u}$ and $I_{12_C} = 2 \times 10^6 \, \text{Hz}$ at $E_{amp0} = 29.3 \, \text{kV/cm}$ and $E_{amp1} = 28.7 \, \text{kV/cm}$

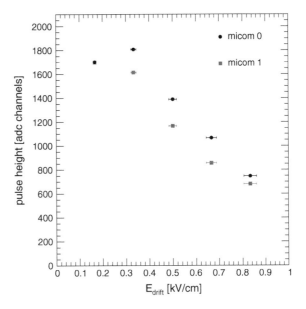

pulse height increases due to an improved separation of ionization charge and a decreasing signal rise time. Due to the constant integration time of the APV25 chips, a decreasing signal rise time leads to an increase of the pulse height. A maximum is reached for $E_{drift} \sim 0.25 \, \text{kV/cm}$. For further increasing drift field, the pulse height decreases again, due to an increasing transverse electron diffusion and a changing electric field line configuration, see Sect. 2.6 for a detailed discussion. Both effects lead to a loss of ionization electrons at the mesh.

The very high particle rates available at HIT, allow for a very stringent test of the floating strip Micromegas high-rate capability. Pulse height spectra, measured for 88.83 MeV/u carbon ions at different particle rates in the first floating strip and the resistive strip Micromegas are shown in Fig. 6.25. Concentrating first on the floating strip detector, two peaks are visible in the spectra for lower rates: A dominant peak, corresponding to the most probable energy loss of a single carbon ion and a faint peak corresponding to a pulse height twice as high. The entries at the lower edge of the spectrum are due to electronics noise and partially detected particle hits. The second, faint peak is created by an accidental merging of signals from two ions into a single hit cluster. This second peak evolves into a flat shoulder with increasing rate. Up to the highest rate the dominant peak is visible, i.e. single particle track reconstruction is possible. An overall pulse height decrease of only 20 % is observed, that is discussed in the following.

The overall current between mesh and anode strips is on the order of $10 \, \mu\text{A}$ at the highest rate. Assuming equal currents on the 55 dominantly hit strips, this corresponds to a recharge current of 180 nA per strip. The current leads to a voltage drop on the hit strips of about 4 V due to the 22 MΩ strip recharge resistor. This corresponds to an amplification field decrease of 0.27 kV/cm.

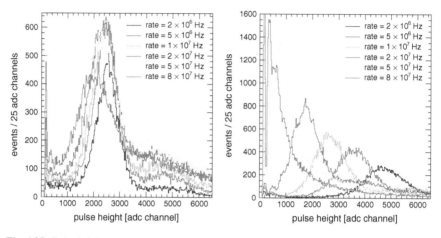

Fig. 6.25 Pulse height spectra, measured with 88.83 MeV/u carbon ions at different particle rates in the floating strip Micromegas 0 (*left*, $E_{amp0} = 30.7$ kV/cm, $E_{drift0} = 0.33$ kV/cm) and the resistive strip Micromegas 2 (*right*, $E_{amp2} = 31.3$ kV/cm and $E_{drift2} = 0.57$ kV/cm)

The gas gain decreases, according to Fig. 6.23, by about 15 % for a field decrease of this size. It seems that, additional to the voltage drop, ion space charge effects in the amplification region and backflow of positive ions from gas amplification into the drift region lead to a small decrease of the measured pulse height on the order of 5 %. The ion backflow can be determined by comparing the measured currents at cathode and mesh. (1.7 ± 0.2)% of the produced positive ions are not neutralized at the mesh but move into the drift region. Space charge effects in the amplification region are assumed to be the dominant component.

In the resistive strip Micromegas, the pulse height decreases by 90 % when going from the lowest to the highest rate. This is caused by the charge up of the resistive strip anode, due to the high strip resistivity. The pulse height decrease could in principle be reduced by increasing the anode high-voltage, although care must be taken, that the non-irradiated detector regions do not enter continuous discharge regimes.

A similar approximation as above for the resistive strip Micromegas is difficult, as the strip resistivity of around 90 MΩ/cm is only coarsely known (Teixeira 2012). The mesh to anode current is of the order of 4 μA. Due to the higher strip pitch, we assume that 110 strips are dominantly hit. The recharge current of 36 nA per strip then leads to a voltage decrease of around 14.5 V, corresponding to an amplification field decrease of 1.1 kV/cm. Using the dependence of the first Townsend coefficient on the electric field, Fig. 2.6, this corresponds to a gain reduction on the order of 40 %. The ion backflow in the resistive strip Micromegas can be determined as above by comparing the currents on mesh and cathode. It is with (12.8 ± 1.0)% considerably higher than in the floating strip detectors. This is on the one hand due to a larger drift field in the resistive strip detector and on the other caused by a smaller optical and electrical opacity of the micro-mesh, consisting of wires with only 18 μm diameter. The combination of both effects may lead to the observed pulse height reduction.

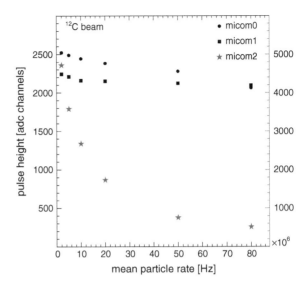

Fig. 6.26 Most probable pulse height as a function of the beam intensity, measured with 88.83 MeV/u carbon ions. The pulse height of the resistive strip Micromegas 2 has been superimposed, its scale is given by the second axis. The electric field parameters were $E_{amp0} = 30.7$ kV/cm, $E_{amp1} = 29.9$ kV/cm and $E_{drift0,1} = 0.33$ kV/cm for the floating strip detectors and $E_{amp2} = 31.3$ kV/cm and $E_{drift2} = 0.57$ kV/cm for the resistive strip detector

An additional contribution due to charge-up of the insulating material into which the resistive strips are embedded is also possible.

The ion backflow in the resistive strip Micromegas is considerably larger than found in a dedicated measurement with a similar detector, presented in Sect. 5.4. This is due to the considerable smaller amplification field during the measurements at HIT. The ion backflow is determined by the ratio of amplification and drift field, as discussed by Colas (2004).

The most probable pulse height, measured with 88.83 MeV/u carbon ions in all three detectors, is shown in Fig. 6.26 as a function of the particle rate. The considerable decrease of the pulse height in the resistive strip detector is visible.

As it has been discussed in Sect. 6.2.2, single particle tracking in proton beams becomes increasingly difficult with increasing beam intensity, due to the high detector occupancy. Signals, produced by individual particles, are merged into few but large hit clusters. Instead of discussing the single particle pulse height, the overall charge deposition by a bunch of coincident protons in the floating strip Micromegas is regarded. It can be calculated by adding the pulse height, measured on all hit strips in an event. The distribution of the sum signal is shown in Fig. 6.27 for different beam intensities. The transition from single particle resolution to integration over a set of proton signals is clearly visible.

The most probable energy loss of ^{12}C-ions in 6 mm Ar:CO_2 93:7 vol% has been calculated as a function of the particle energy using Eq. (2.3). The relation between energy loss and ionization yield is given by Eq. (2.4). In Fig. 6.28 the measured pulse height of Micromegas 0 as a function of the beam energy for carbon beams is shown. The ionization yield has been superimposed and scaled to the pulse height at lower beam energies.

Fig. 6.27 Overall charge
deposition spectrum in
Micromegas 0 for different
proton rates, measured with
$E_p = 221.06$ MeV/u at
$E_{amp0} = 32.0$ kV/cm for the
four lower and
$E_{amp0} = 31.3$ kV/cm for the
highest intensity. The drift
field was
$E_{drift0} = 0.33$ kV/cm

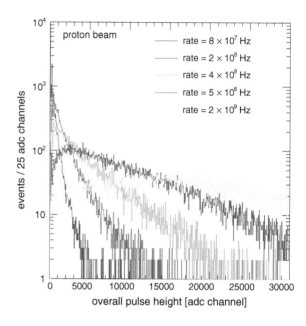

Fig. 6.28 Pulse height in the
first Micromegas as a
function of the particle
energy, measured with
carbon ions at
$I_{12_C} = 5 \times 10^6$ Hz and
electric fields of
$E_{amp0} = 30.7$ kV/cm and
$E_{drift0} = 0.33$ kV/cm.
Superimposed is the
calculated ionization yield.

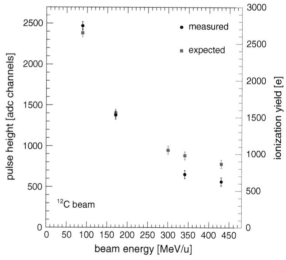

The measured pulse height for higher beam energies is about 30 % lower than
the value expected from the behavior of the ionization yield. The reason for this
deviation is not quite clear. The energy increase is correlated to an increase of the
particle flux density by a factor of 2, due to a decrease of the beam focus FWHM
from 13.4 to 9.8 mm, but the thereby expected pulse height decrease is on the order
of a few percent. It is possible, that the bunch structure of the beam, together with
inhomogeneously filled bunches, leads to considerably higher instantaneous particle

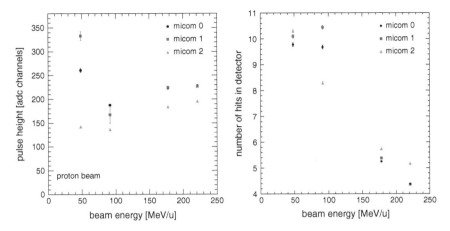

Fig. 6.29 Pulse height (*left*) and mean number of reconstructed hits per event (*right*) in all three Micromegas as a function of the particle energy, measured in proton beams with $I_p = 8 \times 10^7$ Hz. The electric field parameters were $E_{amp0} = 34.6$ kV/cm, $E_{amp1} = 34.1$ kV/cm and $E_{drift0,1} = 0.33$ kV/cm for the floating strip detectors. The resistive strip Micromegas was operated at $E_{drift2} = 0.43$ kV/cm and $E_{amp2} = 39.1$ kV/cm for the two lower and $E_{amp2} = 40.6$ kV/cm for the two upper beam energies

rates. For future measurements, a fast counting detector for high-resolution rate measurements should be included in the tracking system.

In Fig. 6.29 (left) the dependence of the cluster pulse height in all three detectors on the proton beam energy is shown. With increasing energy the pulse height decreases, due to a decreasing proton energy loss. For proton energies $E_p > 100$ MeV/u an increase of the pulse height is observed. This is explained by merging of neighboring clusters into few, but larger hit clusters (Fig. 6.29 (right)), as an energy increase is correlated to a decrease of the beam focus size, which leads to a higher particle flux density. The limitation of the strip detectors with respect to separating individual proton signals becomes clearly visible.

6.2.5 Efficiency

As it has been discussed in Sect. 6.2.3, at least three detector layers are necessary to reliably reconstruct tracks in the high hit-multiplicity environment at HIT. In events with more than one valid hit per event in each detector layer, hits are considered as valid only if they can be matched to a track. Since all detectors are thus involved in hit selection and at the higher rates, there are always several particle signals visible in each detector layer, the single particle efficiency calculation is strongly biased for particle rates above 2 MHz.

Fig. 6.30 Hit efficiency of
the floating strip detectors as
a function of the respective
amplification field. Measured
with 88.83 MeV/u carbon
ions with $I_{12C} = 2 \times 10^6$ Hz
at $E_{\text{drift}} = 0.33$ kV/cm

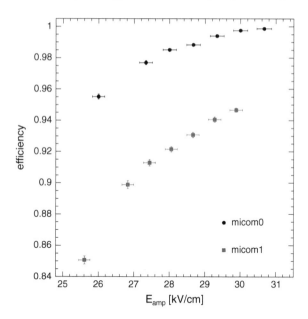

An unbiased single particle detection efficiency could only be determined for the
lower rate measurements with carbon ions. The measured hit efficiency in the two
floating strip Micromegas is shown in Fig. 6.30 as a function of the amplification
field. The lower efficiencies of the second floating strip detector are caused by a
group of ten dead anode strips in the middle of the detector, caused by interrupted
striplines due to low quality etch resist (see Sect. 3.4.2).

The efficiency of Micromegas 0 increases with increasing amplification field
from 0.95 to 0.995, single particle detection efficiencies above 0.98 are reached for
$E_{\text{amp}} \gtrsim 27.5$ kV/cm. For Micromegas 1, similar behavior is observed at a slightly
lower level, due to the ten inefficient strips. Note that due to the large ionization yield
and the diffusion of ionization charge in the drift gap, mesh supporting pillars, that
cover about 1 % of the active region, do not seem to limit the efficiency anymore.

In Fig. 6.31 the detection efficiency of both floating strip detectors is shown as
a function of the drift field. It is almost flat, a minor decrease with increasing drift
field, correlated to a pulse height decrease, Fig. 6.24, is observed.

The efficiency for all three detectors as a function of the carbon ion energy is shown
in Fig. 6.32. For Micromegas 0, efficiencies above 0.995 are observed for all beam
energies. The second floating strip Micromegas 1 shows efficiencies above 0.96, the
resistive strip Micromegas 2 efficiencies above 0.975 for all beam energies. Except
for the dead strip region in the second floating strip detector layer, all Micromegas
show excellent detection efficiencies for carbon ions over the whole energy range.

In Fig. 6.33 the efficiency for the three Micromegas detectors in 80 MHz proton
beams is shown. The measured values are above 0.99 for all energies and show only
a weak energy dependence. Even Micromegas 1 with ten dead strips shows a high

Fig. 6.31 Hit efficiency of the floating strip detectors as a function of the respective drift field. Measured with 88.83 MeV/u carbon ions at $I_{12_C} = 2 \times 10^6$ Hz, $E_{amp0} = 29.3$ kV/cm and $E_{amp1} = 28.7$ kV/cm

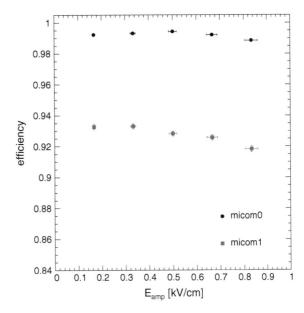

Fig. 6.32 Hit efficiency as a function of the particle energy, measured with carbon beams at a particle rate of 5×10^6 Hz. During carbon ion tracking, the two floating strip Micromegas layers were operated at $E_{amp0} = 30.7$ kV/cm and $E_{amp1} = 29.9$ kV/cm and a common drift voltage of $E_{drift0,1} = 0.33$ kV/cm, the resistive strip detector was at $E_{amp2} = 31.3$ kV/cm and $E_{drift2} = 0.57$ kV/cm

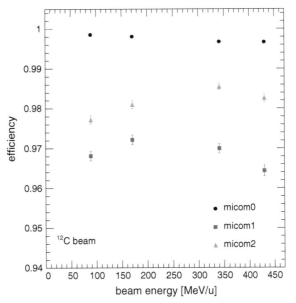

"efficiency", since there is always a particle signal in a properly functioning region of the detector.

The measured hit efficiency for higher rates can not be interpreted as single particle detection efficiency, but is a reliable measure for the overall detector availability, since it represents the fraction of events, in which the detector was able to detect

Fig. 6.33 Hit efficiency as a function of the particle energy, measured with proton beams with a mean particle rate of 80×10^6 Hz. The field parameters were $E_{amp0} = 34.7$ kV/cm, $E_{amp1} = 34.1$ kV/cm, $E_{drift1,2} = 0.33$ kV/cm for the floating strip detectors. The resistive strip detector was operated at $E_{drift2} = 0.43$ kV/cm and $E_{amp2} = 39.1$ kV/cm for the two lower and $E_{amp2} = 40.6$ kV/cm for the two higher energies

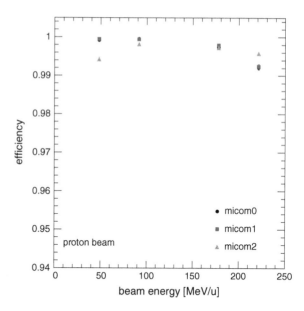

particles. If e.g. discharges would be present, that affect the whole detector, the respective detector would be completely inefficient for single events. The same holds for charge-up effects.

The transition from particle detection efficiency to overall detector availability can be nicely seen in Fig. 6.34 (left). The efficiency of the three Micromegas detectors in carbon beams is shown as a function of the beam intensity. Starting at low intensities, it increases from the particle detection efficiencies, as measured above, to values above 0.99 for $I_{12C} > 20 \times 10^6$ Hz. The availability of the floating strip detectors saturates at 0.998. The values, measured for the resistive strip detector begin to decrease for the highest rates, due to charge-up effects.

Similar behavior is observed for 221.06 MeV/u proton beams as a function of the particle rate. The floating strip detector availability is constantly high, even for the highest rates of 2×10^9 Hz. This demonstrates the good discharge sustaining capabilities of this detector type. Although the discharge rate increases significantly with increasing beam intensity, Sect. 6.2.7, discharges are limited to the affected strips. The overall detector availability and the efficiency of the non-affected strips stays at a very high level.

In the resistive strip Micromegas on the other hand, the overall detector availability decreases significantly with increasing beam intensity, to 0.72 for the highest intensity. This is caused by the overall voltage drop and charge-up effects. The ion back-drift is, due to a smaller drift field, with 5 % considerably smaller than in the discussion in Sect. 6.2.4 and should not have a strong influence on the efficiency.

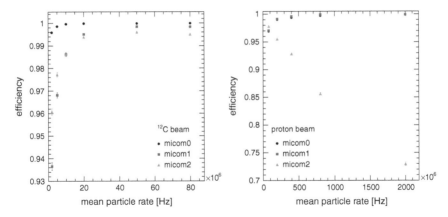

Fig. 6.34 Hit efficiency as a function of the particle rate, measured with carbon beams (*left*) at $E_{12C} = 88.83\,\text{MeV/u}$ and proton beams (*right*) at $E_p = 221.06\,\text{MeV/u}$. During carbon ion tracking, the two floating strip Micromegas layers were operated at $E_{amp0} = 30.7\,\text{kV/cm}$ and $E_{amp1} = 29.9\,\text{kV/cm}$ and a common drift voltage of $E_{drift0,1} = 0.33\,\text{kV/cm}$, the resistive strip detector was at $E_{amp2} = 31.3\,\text{kV/cm}$ and $E_{drift2} = 0.57\,\text{kV/cm}$. In proton beams the amplification voltage of the two floating strip Micromegas layers was reduced in steps of 10 V with increasing particle rate, starting at $U_{amp0} = 510\,\text{V}$ and $U_{amp1} = 550\,\text{V}$, $E_{drift0,1} = 0.33\,\text{kV/cm}$. The resistive strip detector was operated at $E_{drift2} = 0.43\,\text{kV/cm}$ and $E_{amp2} = 39.1\,\text{kV/cm}$ for the lowest and $E_{amp2} = 38.3\,\text{kV/cm}$ for the other rates

6.2.6 Spatial Resolution

The spatial resolution of a certain detector is determined by comparing the predicted particle hit position with the measured hit position in this detector, see also Sect. 4.5. The track reconstruction has been described in Sect. 6.2.3. In the three detector layers are only those hits considered, which could be matched to a track. The predicted track is defined by the two other detector layers. The residual between predicted and measured hit position for many similar tracks follows in principle a Gaussian distribution. Its width is determined by the track accuracy and the spatial resolution of the respective detector. The spatial resolution in this discussion is approximated with the geometric mean method, see Sect. 4.5.2. This method yields correct results, only if the spatial resolution of the tracking detectors is similar, since the method forces similar values for all detectors (Sect. 4.5.4). Due to the small distance of the two, in principle equal, floating strip Micromegas and the symmetric variation of operational parameters in the two floating strip detectors, this assumption is justifiable.

The spatial resolution of the first floating strip detector is shown in Fig. 6.35 as a function of the amplification field. With increasing amplification field, the spatial resolution improves to its optimum value of $\sigma_{SR} = 104 \pm 1\,\mu\text{m}$ at $E_{amp} \sim 29.5\,\text{kV/cm}$. The spatial resolution is almost independent of the amplification field for $E_{amp} \gtrsim 27.5\,\text{kV/cm}$.

Fig. 6.35 Spatial resolution in Micromegas 0 as a function of the amplification field, measured with 88.83 MeV/u carbon ions at $I_{12C} = 2 \times 10^6$ Hz for a constant drift field $E_{drift} = 0.33$ kV/cm

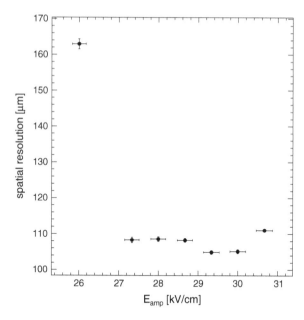

Fig. 6.36 Spatial resolution in Micromegas 0 as a function of the drift field, measured with 88.83 MeV/u carbon ions at the lowest intensity $I_{12C} = 2 \times 10^6$ Hz for $E_{amp} = 29.3$ kV/cm

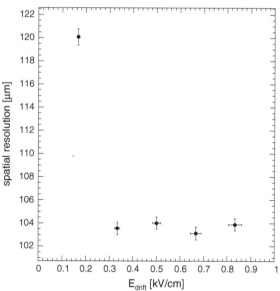

The dependence of the spatial resolution in Micromegas 0 on the drift field is shown in Fig. 6.36. The spatial resolution improves with increasing drift field due to a reduction of the transverse electron diffusion, Fig. 2.5. The transverse diffusion is minimal at $E_{drift} \sim 0.3$ kV/cm. For $E_{drift} \gtrsim 0.3$ kV/cm the spatial resolution depends only weakly on the drift field.

Fig. 6.37 Spatial resolution Micromegas 0 as a function of the beam intensity, measured in 88.83 MeV/u carbon beams at $E_{amp} = 30.7\,kV/cm$ and $E_{drift} = 0.33\,kV/cm$

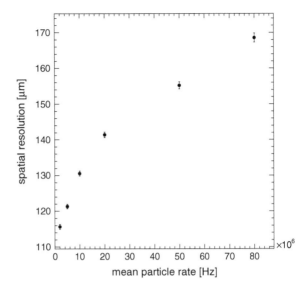

In Fig. 6.37 the spatial resolution is shown as a function of the particle rate. With increasing rate a degradation of the spatial resolution is visible, caused by deterioration of the reconstructed hit position by coincident hits on the same or neighboring strips. For the highest rate still a satisfying resolution of $\sigma_{SR} = (168 \pm 1)\,\mu m$ is observed.

The spatial resolution improves with increasing particle energy from $\sigma_{SR} = (121 \pm 1)\,\mu m$ to $\sigma_{SR} = (97 \pm 1)\,\mu m$, Fig. 6.38. Two effects are responsible for the observed energy dependence: Multiple scattering in the readout structures and the shielding aluminum foil between the two floating strip Micromegas, and a variable cluster width. The cluster width i.e. the number of strips per cluster decreases with decreasing ionization yield, since the charge signals become smaller. It is furthermore influenced by space-charge induced electron diffusion.

Multiple scattering (Sect. 2.2.3) in the readout structure of the first floating strip layer has the strongest influence on the observed spatial resolution. It consist of two layers of 35 μm thick copper strips and a 125 μm thick FR4 board. The radiation lengths and densities of copper and the FR4 board, approximated by the values for epoxy, are given by the Particle Data Group[4] and Gupta (2013). The resulting total radiation length of the board can be calculated with Eq. (2.17) and is $X_0 = 15.1\,g/cm^2$.

Scattering in the shielding aluminum foil is neglected due to the small thickness of $(28 \pm 2)\,\mu m$ and the larger aluminum radiation length. Due to the small lever arm, scattering in the second readout structure will only have a minor influence on the observed spatial resolution and is equally neglected.

---

[4]http://pdg.lbl.gov/AtomicNuclearProperties.

Fig. 6.38 Spatial resolution of the first floating strip Micromegas as a function of the particle energy, measured in carbon beams with $I_{^{12}C} = 5 \times 10^6$ Hz at $E_{amp} = 30.7$ kV/cm and $E_{drift} = 0.33$ kV/cm

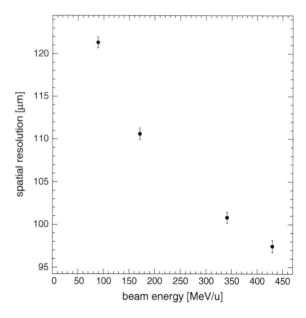

Fig. 6.39 Calculated spread of a narrow pencil beam of carbon ions after 25 mm, induced by multiple scattering in the readout structure of the first floating strip Micromegas layer. Shown is the standard deviation of the resulting Gaussian beam profile

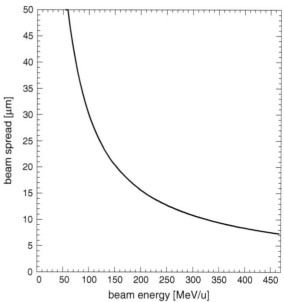

In Fig. 6.39 the spread σ_{ms} (Eq. 2.16) of an initially narrow carbon beam is shown as a function of the beam energy. In a detector with an intrinsic spatial resolution $\sigma_{SR,int}$ on the order of a few μm, the spread σ_{ms} would be equal to the measured

spatial resolution $\sigma_{SR,obs}$. For the floating strip Micromegas, the observed spatial resolution is given by

$$\sigma_{SR,obs} = \sqrt{\sigma_{SR,int}^2 + \sigma_{ms}^2} \, . \tag{6.7}$$

For the lowest energy, the spread σ_{ms} is on the order of 34 μm, for the highest beam energy, it reduces to 8 μm. For an intrinsic spatial resolution on the order of 100 μm, multiple scattering distorts the observed spatial resolution by less than 6 μm.

The dependence of the measured spatial resolution on the beam energy is thus created by a small contribution from multiple scattering and a larger contribution that is possibly correlated to the decreasing ionization density: With increasing particle energy, the ionization yield decreases, such that the tails of the ionization charge distribution are not detected, leading to a smaller and apparently favorable cluster width.

Intentionally shifting the assumed detector position allows for investigating the quality of the track reconstruction method and thus the reliability of the measured spatial resolution. The residual distribution, i.e. the distribution of differences Δx between the predicted and the measured hit position in the first floating strip layer, is shown in Fig. 6.40 for two different particle rates. The detector has been shifted by -0.5 mm in x-direction, the mean residual, measured at the lower rate, shifts accordingly to $+0.5$ mm. This shows, that the observed Gaussian residual distributions represent actually the differences between assumed and measured particle hit positions and are not artificially created by the track reconstruction algorithm, Sect. 6.2.3.

We observe furthermore, that the Gaussian peak sits on an approximately rectangular background, that is caused by mis-measured hit positions, coincidentally selected by the track reconstruction algorithm. The background i.e. the fraction of

Fig. 6.40 Distribution of residuals in Micromegas 0, measured with 88.83 MeV/u carbon beams at two different beam intensities. A fit with the sum of a Gaussian function and a constant (*blue line*) is used to estimate the fraction of well reconstructed tracks

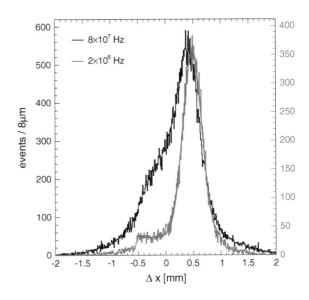

Fig. 6.41 Fraction of well reconstructed tracks with respect to all reconstructed tracks. The number of reconstructed tracks per event is also rate dependent and is shown in Fig. 6.22

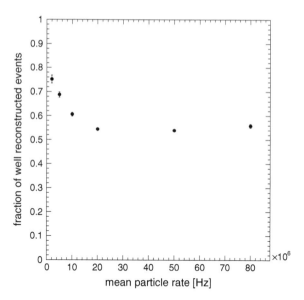

mis-reconstructed events, increases with increasing particle rate, as expected. Even at the highest carbon ion rate of 80 MHz though, the shifted residual peak is still visible, showing that a single particle track reconstruction is still possible.

In order to estimate the fraction of well reconstructed events, the residual distributions can be fitted with a Gaussian function with constant offset. The latter can be used to estimate the number of mis-reconstructed events. Note that this method overestimates the number of mis-reconstructed tracks. The fraction of well reconstructed tracks with respect to all reconstructed tracks is shown in Fig. 6.41. It decreases from 75 % at 2 MHz particle rate with increasing rate and saturates at 50 % for rates above 20 MHz.

6.2.7 Discharge Behavior

The discharge behavior of the floating strip Micromegas detectors in carbon and proton beams is discussed in the following section. The underlying discharge mechanism and the development of streamers has been discussed in Sect. 2.7. Discharges have been monitored (Sect. 6.2.1) by detecting the recharge signal at the common strip high-voltage distributor of each detector. After amplification and discrimination, the discharges were counted with a scaler, read out every 10 s. Due to this large counting period and the inability to measure the instantaneous particle rate time resolved, the determined discharge rates and probabilities are afflicted with considerable errors. The variable spill-break structure of the particle beam (Sect. 6.2.2) makes a correction difficult. Nevertheless, clear trends are visible.

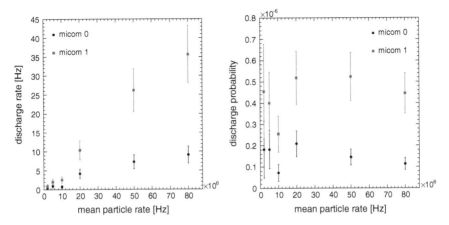

Fig. 6.42 Discharge rate (*left*) and discharge probability per incident particle (*right*) as a function of the beam intensity, measured in the floating strip Micromegas with 88.83 MeV/u carbon ions at $E_{amp0} = 30.7$ kV/cm, $E_{amp1} = 29.9$ kV/cm and $E_{drift0,1} = 0.33$ kV/cm

Two different quantities are shown in the following: Counting the number of discharges during irradiation yields the discharge rate. The discharge probability per incident particle is determined by comparing the number of discharges to the assumed number of particles during irradiation. We assume, that the requested beam intensity is equal to the actual mean particle rate. The irradiation time, representing the duty cycle of the synchrotron, has been estimated from the mean spill and spill-break duration.

In Fig. 6.42 the measured discharge rate and discharge probability in carbon beams is shown as a function of the particle rate. Except for the points at $I_{12C} = 10 \times 10^6$ Hz, the discharge rate increases linearly with the mean particle rate, resulting in an approximately constant discharge probability per incident particle. The low values at $I_{12C} = 10 \times 10^6$ Hz are caused by the mentioned variable delay between consecutive spills.

A linearly increasing discharge rate and thus a constant discharge probability shows, that no cooperative effects between entering particles are present. Cooperative effects are expected, if two particle hit the area in which the streamer develops (0.1 mm×0.1 mm) within the time window of streamer development ($\sim$300 ns). This would correspond to a particle flux density on the order of 30 GHz/cm^2. Even at the highest achievable carbon ion rates, the mean particle flux density of 56 MHz/cm^2 is by three orders of magnitude smaller. A considerable difference between the discharge rates in the first and the second layer of the floating strip Micromegas doublet is observed. As this difference is not observed in protons beams, see Fig. 6.43, the enhanced discharge probabilities in the second layer can be explained by breakup of carbon nuclei and production of strongly ionizing low energetic nuclear fragments, that trigger discharges more easily.

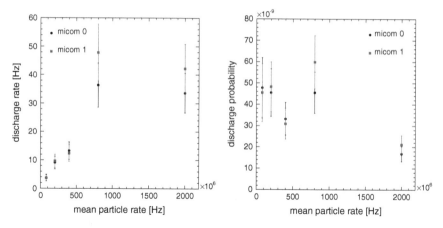

Fig. 6.43 Discharge rate (*left*) and discharge probability per incident particle (*right*) as a function of the beam intensity, measured in the floating strip Micromegas with 221.06 MeV/u protons at $E_{amp0} = 32.0$ kV/cm and $E_{amp1} = 31.7$ kV/cm for the four lower rates and $E_{amp0} = 31.3$ kV/cm and $E_{amp1} = 31.1$ kV/cm for the highest rate. The drift field was at $E_{drift0,1} = 0.33$ kV/cm for all measurements

The measured discharge rates and probabilities in proton beams are shown in Fig. 6.43 as a function of the mean particle rate.

As above, a linear dependence of the discharge rate and thus a constant discharge probability is observed. At the highest rate a lower amplification voltage was set, such that the amplification fields are reduced by 0.67 and 0.61 kV/cm, yielding smaller discharge rates and probabilities. Mean particle rates of 0.8 and 2.0 GHz correspond to mean particle flux densities of 1.6 and 3.9 GHz/cm². These are at least one magnitude smaller than the flux density for cooperative effects, such that the observed independence of the discharge probability on the particle rate agrees with the expectation.

The factor of 10 smaller discharge probabilities as compared to carbon beams are due to the factor 36 smaller energy loss of protons at equal energy per nucleon, see Eq. (2.1). Furthermore, the values shown have been measured at a proton energy of 221.06 MeV/u, compared to 88.83 MeV/u for carbon ions, resulting in a factor of 2 lower mean energy loss. Using higher amplification fields for protons i.e. working at a gas gain, increased by a factor of 2.5, increases the observed discharge probabilities, see Fig. C.10 in the Appendix C.2. The discussion shows, that the occurrence of particle-induced discharges is correlated but not directly proportional to the magnitude of the mean energy loss, as very rare ionization processes in the far tail of the energy loss distribution are responsible for discharges.

The approximate discharge probabilities for different beam energies and amplification fields can be found in the Appendix C.2. No significant dependence on the particle energy is observed, discharge rates increase with increasing amplification fields to maximum values of $4.5 \pm 1.0 \times 10^{-7}$.

In a subsequent measurement, the presented results can be improved by improving the temporal resolution of discharge counting and recording the instantaneous particle rate.

6.2.8 Summary and Outlook

A tracking system consisting of a floating strip Micromegas doublet with low material budget and a resistive strip Micromegas has been successfully tested in high-rate carbon ion and proton beams at the Heidelberg Ion Therapy center. The data acquisition system has been triggered by signals of secondary photons and particles in scintillation detectors. Due to the processing and transfer time of the acquisition system, a readout rate on the order of 1 kHz has been achieved.

A direct measurement of the bunch and the spill structure of proton and carbon ion beams was possible with the Micromegas. The determined bunch spacing and the overall beam structure is in agreement with the expectation.

In carbon beams the reconstruction of on average up to 5.5 particle hits per event and detector layer was possible. The multi-hit resolution is determined by the strip readout structure, the observed resolution is in agreement with the expectation. A Hough transform based track reconstruction algorithm was applied. It allows for the simultaneous reconstruction of several tracks per event. On average 3.2 tracks were found by the algorithm.

The determined spatial resolution of the floating strip Micromegas is better than 180 µm for all carbon ion energies and rates.

The pulse height of charge signals in the floating strip Micromegas shows in carbon beams a rate dependent decrease of 20%, when ramping from the lowest rate of 2 MHz to the highest rate of 80 MHz. In the resistive strip Micromegas a pulse height decrease of 90% is observed. Due to the high rates in proton beams, a separation of single particles is possible at particle rates below 100 MHz. For higher rates, proton hits are merged into few but large clusters.

In the floating strip and the resistive strip Micromegas single particle detection efficiencies close to 1 are observed. At very high rates on the order of several 100 MHz, no sign of a discharge induced efficiency drop is observed in the floating strip detectors. The resistive strip Micromegas suffers from charge-up, such that the detector availability reduces to 73% at the highest proton rate of 2 GHz.

The detectors ran stably at all rates. In both ion beams a rate-independent discharge probability per incident particle on the order of 5×10^{-7} was observed.

The test measurements demonstrate the suitability of floating strip Micromegas detectors for high-rate ion tracking but also reveal the obvious limitations, introduced by the strip readout structures.

In a future application, the detectors are foreseen for ion tracking in imaging applications in ion beam therapy. Their suitability has been demonstrated with the presented measurements. With the current detectors, full tracking at particle flux densities below 7 MHz/cm^2 is possible. A further increase of the multi-hit resolution is possible by using an alternative gas mixture and a faster readout electronics. If a tracking of all particles at particle rates on the order of 100 MHz is desired, pixel readout structures have to be used.

A different, high-rate capable readout system has to be used though, if a high data collection efficiency is desired.

It is expected, that the rate limiting factor in a detection system for ion transmission applications, based on single particle tracking, consisting of Micromegas tracking detectors and a range telescope, constructed with homogeneous scintillator planes, will be the range telescope. The achievable rate capability of the Micromegas detectors is thus sufficient.

References

Biagi SF (1999) Monte Carlo simulation of electron drift and diffusion in counting gases under the influence of electric and magnetic fields. Nucl Instr Meth A 421(12): 234–240. doi:10.1016/S0168-9002(98)01233-9. http://www.sciencedirect.com/science/article/pii/S0168900298012339; ISSN 0168-9002

CAEN S.p.A. (2013a) Technical information manual mod. A1821 Ser Rev 3. http://www.caen.it

CAEN S.p.A. (2013b) SY5527—SY5527LC power supply system, 7 edn. http://www.caen.it

Colas P, Giomataris I, Lepeltier V (2004) Ion backflow in the micromegas TPC for the future linear collider. Nucl Instrum Methods A 535(12):226–230.10.1016/j.nima.2004.07.274. http://www.sciencedirect.com/science/article/pii/S0168900204016080; ISSN 0168–9002 (Proceedings of the 10th International Vienna Conference on Instrumentation)

Feldmeier E, Haberer T, Galonska M, Cee R, Scheloske S, Peters, A (2012) The first magnetic field control (B-train) to optimize the duty cycle of a synchrotron in clinical operation. In: Proceedings of IPAC2012, New Orleans

Gupta M (2013) Calculation of radiation length in materials. PH-EP-Tech-Note-2010-013

Haberer Th, Becher W, Schardt D, Kraft G (1993) Magnetic scanning system for heavy ion therapy. Nucl Instrum Methods A 330(12):296–305.10.1016/0168-9002(93)91335-K. http://www.sciencedirect.com/science/article/pii/016890029391335K; ISSN 0168-9002

Hamamatsu Photonics K.K. (2010) Photomultiplier tube R4124. https://www.hamamatsu.com/

Ondreka D, Weinrich U (2008) The Heidelberg Ion Therapy (HIT) accelerator coming into operation. In: Proceedings of the EPAC08, Genoa

Rinaldi I (2014) Heidelberg University Hospital and LMU Munich, private communication

Schömers C (2014) HIT Betriebs GmbH, private communication

Teixeira A (2012) CERN, private communication

Ziegler JF, Ziegler MD, Biersack JP (2010) SRIM—the stopping and range of ions in matter. Nucl Instrum Methods B 268(1112):1818–1823. doi:10.1016/j.nimb.2010.02.091. http://www.sciencedirect.com/science/article/pii/S0168583X10001862; ISSN 0168–583X (19th International Conference on Ion Beam Analysis)

Chapter 7
Performance and Properties of Micromegas

Floating strip Micromegas are discharge tolerant, high-rate capable, high-resolution micro-pattern gaseous detectors. The properties and the performance of Micromegas detectors have been evaluated in several measurement campaigns and in different applications. In the following, results from the different campaigns, that are discussed in detail in Chaps. 5 and 6, are compared and discussed. Results acquired in the same campaign are displayed with equal color coding in the figures but have not necessarily been obtained with equal operational parameters.

Determination and behavior of the gas amplification is presented in Sect. 7.1. The dependance of the pulse height on the electric drift field is discussed in Sect. 7.2 and compared to the expected behavior of the mesh transparency. In Sect. 7.3 the behavior of the detection efficiency is presented. The track inclination reconstruction in a single detector plane is discussed in Sect. 7.4. In Sect. 7.5 the achievable spatial resolution in different Micromegas detectors and with different particle beams is reported. The discharge behavior and the discharge tolerance of floating strip Micromegas is described in Sect. 7.6. Further improvements of the detectors, the operational parameters and the readout electronics are shortly touched in Sect. 7.8.

7.1 Gas Gain

The typical ionization yield of minimum ionizing particles in Micromegas detectors is on the order of $50\,e$, assuming a drift gap width of $6\,mm$ and operation with an Ar:CO_2 93:7 vol% gas mixture at atmospheric pressure (Sect. 2.2). In order to reliably detect the passage of such a particle, the ionization charge needs amplification by a factor on the order of 10^3 in an avalanche like process in the high-field region between micro-mesh and anode structure. As discussed in Sect. 2.4, the gas gain can be determined with two different methods:

1. In low rate irradiation with charged particles or photons, a determination of the pulse height of charge signals allows for an extraction of the gas gain. The

© Springer International Publishing Switzerland 2015

J. Bortfeldt, *The Floating Strip Micromegas Detector*, Springer Theses,
DOI 10.1007/978-3-319-18893-5_7

ionization yield, charge-to-voltage conversion factors of the preamplifier electronics and detector capacitance dependent factors have to be known (Eq. 2.24).
2. High-rate particle or photon beams can be used to produce a constant current of ionization charge in the drift region. By measuring the current between mesh and anode and comparing it to the known particle or photon rate and the ionization yield, the gas gain can be calculated (Eq. 2.25).

In the measurement campaigns that were presented in this thesis, the absolute gas gain has been determined with the second method. Gas gains between 600 and 2300 for amplification fields between $E_{amp} = 33$ and 36 kV/cm have been measured in a 6.4×6.4 cm^2 floating strip Micromegas with an amplification gap width of (160 ± 3) μm under lateral irradiation with 20 MeV protons (Sect. 5.3). Gas gains above 1800 were sufficient to detect cosmic muons, traversing a 15 mm wide drift region of the detector.

In 20 MeV proton tracking measurements approximate gas gains could be determined in the two layers of a low material budget floating strip Micromegas doublet (Sect. 6.1). Due to the short ionization paths of only 6 mm, the measured currents were relatively low, such that the determined gas gains are afflicted with an uncertainty of at least 25 %. Another source of uncertainty comes from the determination of the mean energy loss of protons.

For the Micromegas layer with an amplification gap width of (150 ± 1) μm, gas gains between 500 and 8000 were measured for amplification fields between $E_{amp} = 31$ and 37 kV/cm. For the second layer with an amplification gap width of (164 ± 1) μm, amplification factors between 700 and 6000 were observed for amplification fields between $E_{amp} = 31$ and 35 kV/cm.

When taking the different gas pressures in the Micromegas into account, gas gains found in proton tracking and in cosmic muon tracking under proton irradiation are in agreement for equal amplification fields within their respective uncertainties. Using the gas gain parametrization Eq. (2.23), a gain reduction on the order of 30 % for a pressure increase of 35 mbar is expected.

The Micromegas characterization setup, that is described in detail in Bortfeldt (2010, Chap. 4), has been used to determine the gas gain of a standard Micromegas under controlled pressure and temperature conditions. Signal shapes for 5.9 keV X-rays, emitted by a ^{55}Fe-source in a 9×10 cm^2 standard Micromegas, read out with a single charge sensitive preamplifier, were acquired with a 12 bit 1 GHz flash ADC with a recording time window of 2520 ns (CAEN S.p.A. 2010). The detector was operated in a Memmert ICH 256 climate chamber, allowing for investigating the detector performance at stabilized temperatures between -10 and 35 °C.

The gas gain has been determined by recording the signal pulse height with a charge sensitive preamplifier (method 1). The temperature and pressure dependance of the gas gain has been investigated in depth, leading to the parametrization coefficients, stated in Sect. 2.4 (Lippert 2012). The same setup has been used in measurement by Kuger (2013) and the author, used to verify an advanced Micromegas GARFIELD simulation. The simulation is able to correctly predict gas gains in Micromegas with a precision on the order of 15 %, by including Penning and

Penning-like energy transfers in gas avalanches. For a standard and a resistive strip Micromegas with 128 μm amplification gap, operated with an Ar:CO_2 93:7 vol% gas mixture at 1013 mbar and 20 °C, gas gains between 400 and 5000 were observed in measurements and in simulation for amplification fields between E_{amp} = 36 and 43 kV/cm (Kuger 2013).

The dependance of the pulse height on the amplification field has been investigated in most measurement campaigns, presented in this thesis. Since the pulse height is, for a fixed drift field, directly proportional to the gas gain, the relative behavior of the gas gain has been investigated. In all measurements the expected exponential behavior (Fig. 2.6) was observed.

7.2 Mesh Transparency

The electric drift field dependance of the pulse height is often subsumed under the term mesh transparency. This is not entirely correct, as the electron mesh transparency describes only the transmission of electrons, produced in ionization processes, into the high field region between mesh and anode structure. The electron transparency of micro-meshes in Micromegas is considerably larger than the optical transparency. This is due to the high amplification field, that acts like a funnel on ionization electrons, entering from the drift into the amplification region. The electron mesh transparency decreases with increasing drift field due to two effects:

1. With increasing drift field, the configuration of electric field lines in the drift region changes: More field lines end on the mesh, instead of reaching into the amplification region. Thus more electrons are lost on the mesh.
2. Due to the considerable transverse diffusion of electrons, electrons do not follow strictly the electric field lines. An increased transverse diffusion enhances the mesh opacity as electrons diffuse out of the funnel and can be absorbed by the mesh.

Since the transverse diffusion in Ar:CO_2 gas mixtures is minimal for low fields, where also the electric field line configuration is optimal, the two effects add up, leading to the typical pulse height decrease with increasing drift field. The simulated drift field dependance, see Sect. 2.6 and Fig. 2.8, is caused by these two effects. Maximum mesh transparency is reached for small fields.

Starting at low drift fields, an initial increase of the pulse height with the drift field is often observed. This can be understood as follows:

1. The rise time of charge signals, produced by traversing charge particles, is given by the sum of the maximum drift time of electrons from cathode to mesh and the maximum drift time of positive ions from anode to mesh, which is on the order of 150 ns. The pulse height depends on the charge fraction, that arrives on the readout strips during the constant integration time of the applied preamplifier electronics. If the signal rise time is longer than the preamplifier integration time, smaller

Fig. 7.1 Maximum electron drift time as a function of the drift field. Determined with a MAGBOLTZ simulation for two different Ar:CO$_2$ gas mixtures at 20 °C and 1013 mbar and two different drift gap widths

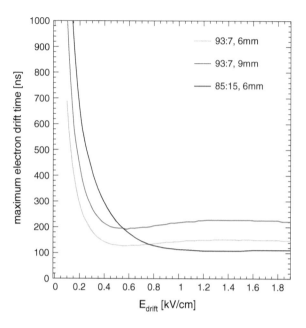

pulse heights are observed. The maximum electron drift time in Micromegas with two different drift gap widths, operated with an Ar:CO$_2$ gas mixture, is shown in Fig. 7.1. A larger maximum electron drift time and thus signal rise time is expected for a detector with larger drift gap. This dependency is the dominant contribution to the observed initial rise.

2. A minimum electric field is necessary to reliably separate electrons and positive ions, produced in ionization processes. An increasing drift field improves the separation and decreases attachment.

A comparison of the simulated with the measured drift field dependence of the pulse height is shown in Fig. 7.2 for measurements with an Ar:CO$_2$ 93:7 vol% gas mixture. The shown pulse heights are scaled to the simulated mesh transparency for large drift fields, as best agreement is expected in this region. The data have been acquired with a 48 × 50 cm^2 floating strip Micromegas in 120 GeV pion beams (Sect. 5.2) and a 9 × 10 cm^2 standard Micromegas in 90 GeV muon beams (Sect. 5.1). Signals of the floating strip Micromegas with a 9 mm wide drift region were acquired with APV25 based electronics, which features a typical shaping and integration time of 50 ns. The small standard Micromegas with a drift gap width of 6 mm was read out with Gassiplex based front-end boards with a signal shaping time of 650 ns.

Starting at low drift fields, the measured scaled pulse heights rise and approach the calculated values, due to a decreasing signal rise time. For $E_{\text{drift}} \gtrsim 0.5$ kV/cm and $E_{\text{drift}} \gtrsim 0.8$ kV/cm the behavior of the measured values in the standard and the floating strip Micromegas, respectively, is consistent with the simulation.

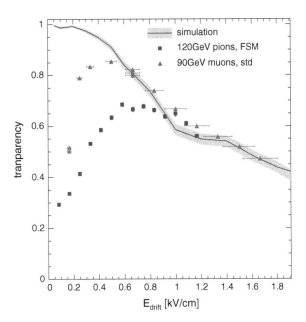

Fig. 7.2 Electron mesh transparency as a function of the drift field in Micromegas with woven meshes with 18 μm diameter wires for an Ar:CO$_2$ 93:7 vol% gas mixture. Superimposed are GARFIELD simulation results and measured scaled pulse heights in the 48 × 50 cm^2 floating strip Micromegas (FSM) in 120 GeV pion beams and in a 9 × 10 cm^2 standard Micromegas (std) in 90 GeV muon beams

The difference between the two curves is rooted in the larger drift region of the floating strip Micromegas and the shorter integration time of the applied APV25 chip, as compared to the Gassiplex preamplifier.

Figure 7.3 shows simulated electron mesh transparencies in an Ar:CO$_2$ 85:15 vol% gas mixture and scaled pulse heights, measured in a 9 × 10 cm^2 standard Micromegas in 120 GeV pion and 90 GeV muon beams.

As already discussed in Sect. 5.1, the behavior in muon and pion beams is equivalent. The initial rise of the measured pulse height is as above due to a decreasing signal rise time with increasing drift field. The behavior of scaled pulse height and simulated transparency agrees for $E_{drift} \gtrsim 0.8$ kV/cm. Maximum mesh transparency is observed at a higher field than in measurements with the 93:7 vol% gas mixture. The differing behavior can be correlated to the differing transverse electron diffusion, Fig. 2.5, and signal rise time, Fig. 7.1.

In a cosmic muon detecting floating strip Micromegas under lateral irradiation with 20 MeV protons (Sect. 5.3), the electron mesh transparency has been determined with an alternative method: Due to the sufficiently large ionization yield of 20 MeV protons and a large gas gain, the measurement of the current between mesh and anode structure allowed for an investigation of the mesh transparency, independent of integration time effects of the preamplifier electronics. This opens a window for comparison of the detector performance with simulation at low drift fields.

Due to the smaller optical transparency of a micro-mesh consisting of 25 μm diameter wires, the drift field dependence of the transparency in detectors with this kind of mesh is stronger. In Fig. 7.4 simulated and measured electron mesh transparency are superimposed. As before, the measured data has been scaled to the simulation.

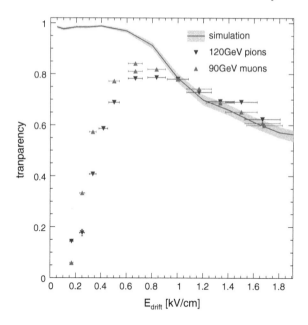

Fig. 7.3 Electron mesh transparency as a function of the drift field in Micromegas with woven meshes, consisting of 18 μm diameter wires for an Ar:CO$_2$ 85:15 vol% gas mixture. Shown are GARFIELD simulation results and measured scaled pulse heights in a $9 \times 10\,\text{cm}^2$ standard Micromegas in 120 GeV pion and 90 GeV muon beams

The measured behavior of the electron mesh transparency, determined from a direct measurement of the current between mesh and anode, agrees well with the simulated values in the investigated range $0.13\,\text{kV/cm} \leq E_{\text{drift}} \leq 0.4\,\text{kV/cm}$. The pulse height, measured in a floating strip Micromegas doublet with APV25 based readout electronics in 20 MeV proton beams (Sect. 6.1), shows the expected initial rise and agrees with the simulated mesh transparency for $E_{\text{drift}} \geq 0.4\,\text{kV/cm}$. This demonstrates, that the initial rise of the pulse height, usually observed with Gassiplex or APV25 readout electronics, is dominantly caused by a decrease of the signal rise time.

The discussion shows, that in order to reach maximum pulse height with a specific detector, interfaced with a specific readout electronics, the drift field dependence of the pulse height should be investigated experimentally. It is not sufficient to rely on previously known optimum drift field values, although one can generally say, that optimum values will be of the order 0.5 kV/cm. If possible, the integration and shaping time of the applied readout electronics should be adapted to the signal rise time in the interfaced Micromegas detector.

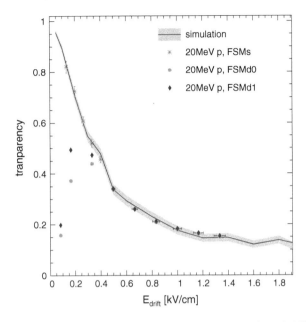

Fig. 7.4 Electron mesh transparency as a function of the drift field in floating strip Micromegas with woven meshes, consisting of 25 μm diameter wires for an Ar:CO$_2$ 93:7 vol% gas mixture. Shown are GARFIELD simulation results (*black line*), measurements of the current between mesh and anode in a $6.4 \times 6.4 \, cm^2$ floating strip Micromegas under lateral irradiation with 20 MeV protons (*dark green stars*) and measured pulse heights in the two layers of a $6.4 \times 6.4 \, cm^2$ floating strip Micromegas doublet with low material budget in 20 MeV proton beams (*yellow circles* and *blue diamonds*). Measured data is scaled to the simulation results

7.3 Detection Efficiency

The detection efficiency describes the capability of a detector to register the passage of traversing particles. In tracking applications, a high detection efficiency for the respective radiation is desired. The efficiency is limited by the pulse height of the charge signals and global and local dead time effects. The latter are caused by discharges and charge-up. Furthermore, the mesh supportive pillars, that cover about 1–2 % of the active area, lead to inefficient spots in the Micromegas. This defines an upper limit for the detection efficiency for perpendicularly incident minimum ionizing particles.

The hit efficiency of a standard Micromegas for high-energy pions and muons (Sect. 5.1) is shown in Fig. 7.5. The detectors were operated with an Ar:CO$_2$ 85:15 vol% gas mixture. The efficiency increases with increasing drift field to values above 0.96 due to an increasing pulse height and an additional more subtle effect: For small drift fields, the mesh transparency is high, but the most probable pulse height is low, since the signal rise time is long and the preamplifier electronics registers only the charge arriving within its integration time. The efficiency is likewise low, first due to

Fig. 7.5 Detection
efficiency as a function of
the drift field, measured with
a $9 \times 10\,cm^2$ standard
Micromegas in pion and
muon beams. The detector
was operated with an
Ar:CO$_2$ 85:15 vol% gas
mixture. Significant jumps in
the efficiency found in pion
beams, are due to dead time,
induced by sporadic
discharges

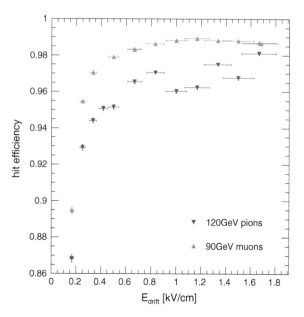

the clustering of ionization charge, that leads to a large straggling of the measured pulse heights and second due to the small most probable pulse height. The straggling of the observed pulse height is enhanced by the unfavorable relation between signal rise time and preamplifier integration time.

The detection efficiency is not directly related to the absolute mesh transparency i.e. pulse height (compare Figs. 7.3 and 7.5): Although the electron mesh transparency and thus the most probable pulse height decreases with increasing drift field, the efficiency remains in the efficiency plateau at values above 0.98 for muons and 0.96 for pions. Due to the larger electron drift velocity, the signal rise time is small. The pulse height straggling is thus smaller than at low fields, and the efficiency remains high.

The 2% difference between the measured efficiencies in pion and muon beams is caused by the elevated discharge rate in the higher-rate pion measurements.

Measured hit efficiencies in three different detectors and in three different particle beams are shown in Fig. 7.6 as a function of the drift field for measurements with an Ar:CO$_2$ 93:7 vol% gas mixture. The data shown has been measured with a $9 \times 10\,cm^2$ standard Micromegas in 90 GeV muon beams (Sect. 5.1) and a low material budget floating strip Micromegas in 20 MeV proton (Sect. 6.1) and 88.83 MeV/u carbon ion beams (Sect. 6.2).

The measured values follow the universal behavior: Starting at low drift fields, the efficiency increases to values above 0.98. As discussed above, this increase is correlated to a mean pulse height increase *and* a lower pulse height straggling due to a decreasing signal rise time. Due to the more homogeneous ionization of low-energy

Fig. 7.6 Efficiency as a function of the drift field for measurements with an Ar:CO$_2$ 93:7 vol% gas mixture. Shown is data measured with a $9 \times 10\,cm^2$ standard Micromegas in 90 GeV muon beams, the second layer of a $6.4 \times 6.4\,cm^2$ floating strip Micromegas doublet in 20 MeV proton beams and the first layer of the same detector doublet in 88.83 MeV/u carbon ion tracking measurements

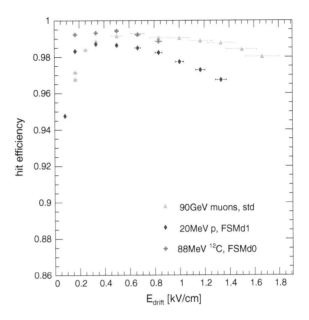

ions and the thus smaller pulse height straggling, the increase is less pronounced in ion beams. For further increasing drift field, the observed efficiency remains at a high level and depends only weakly on the drift field.

Discharges or charge-up effects that affect the whole Micromegas also limit the single particle detection efficiency. In the high-rate proton and carbon ion tracking measurements at the Heidelberg Ion Therapy center (Sect. 6.2), the rate dependent overall detector availability has been measured in floating strip and in resistive strip Micromegas at hit rates between 20 MHz and 2 GHz by counting events, in which not a single particle was detected in the detector. For the two floating strip detectors, the uptime was above 0.99 for all rates. This demonstrates the discharge tolerance of floating strip Micromegas. Particle induced discharges at a rate of several 10 Hz affect only the few involved strips and do not lead to an overall voltage drop and thus inefficiency of the detector. The uptime of the resistive strip Micromegas decreases from above 0.99 at lower rates to below 0.75 at the highest rate, due to considerable charge-up of the resistive strip anode, which is not observed in floating strip Micromegas.

The discussion shows, that high particle detection efficiency for different particle types can be reached in floating strip Micromegas. Except for the initial efficiency increase with increasing drift field, it depends only weakly on the drift field, that can thus be optimized for other parameters like pulse height or spatial resolution.

7.4 Single Plane Angle Reconstruction—μTPC

Micromegas with strip readout structures can measure particle hit positions in one dimension by comparing the pulse height of charge signals on adjacent strips (Sect. 4.3). If additional to the absolute charge on the readout strips, the arrival time of ionization electrons on each readout strip can be measured, which is equivalent to determining the beginning of the charge signal, two-dimensional tracklet reconstruction is possible. With the either known or in situ determined electron drift velocity, the measured drift time of electrons can be translated into a drift length. The readout structure yields hit information in x-direction, the drift time provides additional information in z-direction. For inclined tracks, the z-x-data points can be fitted with a straight line, yielding track inclination and intercept of the tracklet with the readout structure (Sect. 4.8).

This so called μTPC reconstruction method has been investigated and optimized in two test measurements with a $48 \times 50\,\mathrm{cm}^2$ floating strip Micromegas in $120\,\mathrm{GeV}$ pion beams (Sect. 5.2) and a low material budget floating strip Micromegas doublet in $20\,\mathrm{MeV}$ proton beams (Sect. 6.2). The detectors under test have been tilted with respect to the beam direction, such that the relative particle track inclination was known.

The most probable reconstructed angle and the angular resolution as a function of the true track inclination are shown in Fig. 7.7.

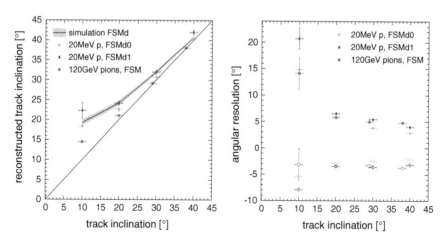

Fig. 7.7 Most probable reconstructed track inclination (*left*) and angular resolution (*right*) as a function of the track inclination. Due to systematic effects, resolutions for angles smaller and larger than the most probable reconstructed angle are different and thus stated separately as negative and positive values. Shown are results determined with a low material budget floating strip Micromegas doublet in $20\,\mathrm{MeV}$ proton beams (FSMd0 and FSMd1) and with the $48 \times 50\,\mathrm{cm}^2$ floating strip Micromegas (FSM) in $120\,\mathrm{GeV}$ pion beams. In the *left* plot, the expected reconstructed angles are superimposed, that were determined with a LTspice simulation, taking the capacitive coupling of adjacent anode strips into account

The most probable reconstructed angle is always larger than the true track inclination and approaches the correct value for increasing track angles. The observed
behavior is caused by the detector specific systematic deviations of the reconstruction
method (Sect. 4.8.4): Capacitive coupling of neighboring strips, a systematic shift of
the measured charge positions on hit strips at the edges of the strip cluster and misidentified noise on adjacent, not actually hit strips, lead to a systematic enlargement
of reconstructed track inclinations.

In Fig. 7.7 results from a LTspice based simulation are superimposed, that predicts
reconstructed track angles in the two layers of the $6.4 \times 6.4 \, cm^2$ floating strip
Micromegas doublet on the basis of the systematic effects. The internal setup of the
simulated detector enters the simulation. Observed and simulated reconstructed track
inclinations agree well within their respective uncertainties.

The deviation between true and reconstructed track angle is smaller in the $48 \times
50 \, cm^2$ floating strip Micromegas. This is caused by the larger ratio between coupling
capacitance of anode- to readout-strips and capacitance of neighboring strips and the
smaller strip pitch of 0.25 mm as compared to 0.5 mm for the small floating strip
detectors.

Due to the systematic deviations and the occasional mis-identification of noise as
hit, the distribution of reconstructed angles shows asymmetric tails towards larger
track angles. The angular resolution is thus stated separately for angles smaller and
larger than the most probable. The angular resolutions for all three detectors are
similar and improve with increasing track inclination from $\left(^{+15}_{-5}\right)^\circ$ at 10° inclination
to $\left(^{+5}_{-4}\right)^\circ$ at 40° inclination. The resolution observed in the small floating strip doublet
detectors at large track inclinations is by 2° better, due to the denser ionization of
20 MeV protons and the larger strip pitch, that avoids charge spreading over many
strips.

The discussion shows, that the reconstruction of track inclination in a single
Micromegas plane is possible for minimum ionizing particles and 20 MeV protons
with track inclinations from 10° to 40°. Systematic effects shift reconstructed angles
towards larger values, these effects can be quantitatively described by a LTspice
based detector simulation. A smaller strip pitch leads to smaller deviations between
reconstructed and true angles at low inclinations, larger strip pitch enables slightly
better angular resolutions at larger track inclinations.

7.5 Spatial Resolution

Spatial resolution is the accuracy of the hit position measurement of traversing particles (Sect. 4.5). It depends on the form of the readout structure, the pulse height of
charge signals on the anode, the electron diffusion in the drift region and on the track
inclination. Optimum spatial resolutions were reported in low-diffusion gas mixtures
by Derré et al. (2001).

In the measurements presented in this thesis, the spatial resolution of a detector under test has been determined by comparing measured hit positions with hit positions, predicted by interpolation or extrapolation of a track, defined by reference detectors. The accuracy of the hit prediction can be calculated from the spatial resolutions of the reference detectors (Eq. B.11), that have beforehand been determined with the geometric-mean method (Sect. 4.5.2)

The distribution of residuals between measured and predicted hit positions for many similar tracks was fitted with a single or double Gaussian function. The standard deviation of the dominant Gaussian allows then for an extraction of the spatial resolution by quadratically subtracting the track accuracy (Eq. 4.16).

The spatial resolution for perpendicularly incident particles, achieved with a $9 \times 10\,\mathrm{cm}^2$ standard Micromegas in 90 GeV muon beams (Sect. 5.1), a $48 \times 50\,\mathrm{cm}^2$ floating strip Micromegas in 120 GeV pion beams (Sect. 5.2) and a low material budget $6.4 \times 6.4\,\mathrm{cm}^2$ floating strip Micromegas in 88.83 MeV carbon ion beams (Sect. 6.2) is shown in Fig. 7.8 as a function of the drift field.

The standard and the large floating strip Micromegas feature an anode strip pitch of 250 µm, the readout structure of the small floating strip Micromegas doublet is formed by anode strips with a pitch of 500 µm.

The observed spatial resolution is optimal at low drift fields $E_{\mathrm{drift}} \sim 0.25\,\mathrm{kV/cm}$. This can be correlated to the minimum of the transverse electron diffusion, Fig. 2.5. In the small floating strip Micromegas an overall larger value for the spatial resolution is observed.

The optimum achievable resolution in the standard Micromegas, read out with the Gassiplex based readout electronics, is on the order of 30 µm. In the large floating strip

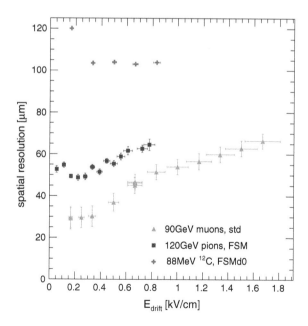

Fig. 7.8 Spatial resolution for perpendicularly incident particles as a function of the drift field for an Ar:CO$_2$ 93:7 vol% gas mixture. Superimposed are optimum spatial resolutions measured with $9 \times 10\,\mathrm{cm}^2$ standard Micromegas in 90 GeV muon beams (*green triangles*), $48 \times 50\,\mathrm{cm}^2$ floating strip Micromegas in 120 GeV pion beams (*red squares*) and a low material budget $6.4 \times 6.4\,\mathrm{cm}^2$ floating strip Micromegas with 0.5 mm strip pitch in 88.83 MeV carbon ion beams (*cyan crosses*)

Micromegas, an optimum resolution of 50 µm is reached, the larger value is correlated to the APV25 readout electronics: The optimum spatial resolution is correlated to low electron diffusion, which is reached at small drift fields. Caused by the short APV25 shaping time and the long signal duration at low drift fields, the pulse height is rather low at small fields, leading together to an overall worse resolution. The optimum resolution of the small floating strip Micromegas with 500 µm strip pitch is a factor of two worse and is at the order of 105 µm.

The track inclination dependence of the spatial resolution has been investigated with the 48×50 cm^2 floating strip Micromegas in 120 GeV pion beams (Sect. 5.2, Fig. 5.30). The accuracy of the hit position, defined by the charge-weighted mean of strips in the hit cluster (Eq. 4.2), degrades significantly with increasing track inclination, a resolution of only (1.15 ± 0.07) mm is observed for tracks with an inclination of 38°.

For inclined tracks, the zero of the µTPC fit to z-x-data points (Sects. 7.4 and 4.8.1) provides an additional particle hit information. The accuracy of the µTPC hit position is better than the accuracy of the usual centroid method for tracks with an inclination above 18°. It is on the order of (0.37 ± 0.04) mm for inclined tracks between 20° and 40°, and shows only a weak dependence on the track inclination in this region. A combination of the two hit positions is in principle possible and can result in better spatial resolutions for inclined tracks.

7.6 Discharge Behavior

Discharges between mesh and anode structure are observed in Micromegas. They are non-destructive and do not seem to permanently alter the detector performance, but due to the necessary recharge of mesh or anode, dead time and thus inefficiency is created. Spontaneous discharges are due to small detector defects and dust in the amplification region. Their occurrence can be reduced by thorough cleaning prior to assembly, assembly under clean room conditions and commissioning with elevated high-voltage under air.

Particle induced discharges are triggered by ionization processes with large energy loss, that lead to charge densities exceeding 2×10^6 e/0.01 mm^2 in Ar:CO$_2$ 93:7 vol% gas mixtures. By operating the detectors at a lower gas gain or using light drift gases with faster ion mobility, the discharge probability can be reduced (Thers et al. 2001).

The occurrence of discharges seems unavoidable in Micromegas, but their influence on the detector efficiency can be greatly reduced by adapting the readout structure. In resistive strip Micromegas, the copper anode strips are covered with a thin insulating layer into which resistive strips with a strip resistivity on the order of several MΩ/cm are embedded (Sect. 3.1.2). In this thesis, an alternative approach with floating copper anode strips has been chosen (Sect. 3.2): Copper anode strips are individually connected to high-voltage via resistors with a typical resistance of several 10 MΩ, signals are capacitively decoupled via small capacitances. In Sect. 3.3,

Table 7.1 Discharge investigation measurements

Detector	Beam	Gas Ar:CO_2 [vol%]	E_{amp} [kV/cm]	P_p
$6.4 \times 6.4 \, cm^2$ floating strip	5.5 MeV alphas	93:7	37	10^{-1}
$9 \times 10 \, cm^2$ standard	120 GeV pions	85:15	41	10^{-5}
$48 \times 50 \, cm^2$ floating strip	120 GeV pions	93:7	38	10^{-4}
$6.4 \times 6.4 \, cm^2$ floating strip	Cosmic muons + 20 MeV protons	93:7	36	3×10^{-7}
$6.4 \times 6.4 \, cm^2$ floating strip	20 MeV protons	93:7	33	2×10^{-7}
$6.4 \times 6.4 \, cm^2$ floating strip	88.83 MeV/u ^{12}C	93:7	30	5×10^{-7}
$6.4 \times 6.4 \, cm^2$ floating strip	221.06 MeV/u protons	93:7	30	5×10^{-8}

Stated are the detector type, the particle type and mean energy, the used gas mixture, typical amplification fields and typical discharge probabilities P_p. In cosmic muon tracking under proton irradiation, the discharge probability per incident proton is given

the good discharge tolerance of floating strip Micromegas could be shown in theory and in measurements.

Discharge probabilities per incident particle depend on the particle type and energy, the ionization yield, mesh transparency and gas gain, the amplification gap width and the applied gas mixture. In this thesis, the discharge probabilities have been investigated in $9 \times 10 \, cm^2$ standard Micromegas in 120 GeV pion beams, using an Ar:CO_2 85:15 vol% gas mixture (Sect. 5.1), in a $48 \times 50 \, cm^2$ floating strip Micromegas in 120 GeV pion beams, using an Ar:CO_2 93:7 vol% gas mixture (Sect. 5.2), in a 6.4×6.4 cosmic muon sensitive floating strip Micromegas under lateral irradiation with 20 MeV protons (Sect. 5.3) and in proton and carbon ion tracking experiments with different ion energies (Sects. 6.1 and 6.2).

Due to the large parameter space covered in the measurement campaigns, Table 7.1 states the order of magnitude of the measured discharge probabilities.

Except for the dedicated discharge measurements in Sect. 3.3, the discharge probabilities in low energy ion irradiation measurements are on the order of 10^{-7}. In medium-rate minimum ionizing particle tracking applications, typical discharge probabilities are on the order of 10^{-4}. The three orders of magnitude larger probabilities are caused by the higher gas gains, needed due to the considerably smaller energy loss of minimum ionizing particles, and the larger energy loss straggling as compared to lower energy ions. Due to the relatively low muon fluxes in the calibration measurements, presented in Sect. 5.1, no direct comparison of discharge probabilities for minimum ionizing muons and pions was possible. It is expected though, that discharge probabilities for hadrons are higher than for muons (Thers et al. 2001), due to the enhanced nuclear interaction cross section for hadrons, thus creating highly ionizing nuclear fragments.

The measurements were performed at different particle rates between 10 Hz and 2 GHz. No rate and particle flux density dependence of the discharge probability has been observed. Cooperative effects between temporal coincident particles are not observed. Discharges in floating strip Micromegas are localized. Cooperative effects at normal particle tracking operational parameters are expected for particle flux densities beyond 30 GHz/cm², which exceed by three orders of magnitude rates at which single particle tracking in Micromegas with strip readout is still possible.

The slightly lower efficiency of standard Micromegas in pion beams as compared to muon beams (Sect. 7.3 and Fig. 7.5) showed the necessity for higher discharge tolerance of Micromegas in high-rate environments. Micromegas with floating strip anode have been developed as a consequence. In Sect. 3.3 the high discharge tolerance of a small floating strip Micromegas has been demonstrated.

This discussion shows, that floating strip Micromegas in various tracking applications are discharge tolerant up to the highest particle rates.

7.7 High-Rate Capability

The rate capability of gas detectors is in general limited by space charge effects due to slowly drifting positive ions, created in gas amplification processes. In order to reach higher maximum rates, the drift time of ions is reduced as much as possible, e.g. by gating grids in wire TPCs or by small and confined gas amplification regions such as the holes of a GEM foil. Micromegas are intrinsically high rate capable, due to short ion drift paths on the order of 100 μm from anode to mesh (Sects. 1.1.1 and 2.5). Only a small fraction of ions on the order of a few percent passes through the mesh holes into the drift region and drifts slowly towards the cathode (Sect. 5.4). The influence of this so called ion backflow on the detector tracking performance has been investigated with flux densities of up to 60 MHz/cm² (Sect. 6.2). Positive ion space charge effects in the amplification region and ion backflow induced pulse height reduction on the order of 5 % are observed. Thus space charge effects are of no concern to the rate capabilities of Micromegas.

The high-rate capability of a Micromegas detector is closely related to discharges and charge-up effects. It has been shown, that discharge probabilities are up to highest rates of 2 GHz independent of the particle rate (Sect. 6.2.7). Dead time is produced by the necessary recharge of mesh or anode structure after a discharge. While in standard Micromegas, the dead time of the whole detector is on the order of 100 ms, it could be reduced in floating strip Micromegas to about 2 ms for the affected strips. This leads to a discharge induced inefficiency, that is by more than three orders of magnitude lower than in standard Micromegas (Sect. 3.2.1).

In floating strip Micromegas a charge-up induced pulse height decrease, that is due to the voltage drop over the strip recharge resistor, of only 15 % has been observed at particle rates of 80 MHz, corresponding to particle flux densities on the order of 60 MHz/cm². The detector ran stably up to rates of 2 GHz, no discharge induced global detector availability drops were observed. Note that single particle tracking is possible with the constructed Micromegas with strip readout up to particle flux densities of $7\,MHz/cm^2$. If single particle tracking at higher rates is desired, the readout structure has to be further segmented into pixels.

Comparing the pulse height reduction at rates of 80 MHz and the behavior of the overall detector availability at rates of 2 GHz in floating strip and in resistive strip Micromegas, demonstrates the suitability of floating Micromegas for very high-rate tracking applications. In resistive strip Micromegas a pulse height decrease by 90 % is observed, in contrast to floating strip Micromegas, where a pulse height decrease of 20 % is measured for a particle rate of 80 MHz. The detector uptime of a resistive strip Micromegas decreases to below 0.75 at a particle rate of 2 GHz, the uptime of the floating strip detectors stays above 0.99.

7.8 Possible Improvements

Tracking systems consisting of floating strip Micromegas, can in several ways be further improved for future measurements. In the following, adaptations of the detector design, detector operation and the readout system are considered. For many applications, the rate capability and the discharge tolerance of the current design is more than sufficient.

7.8.1 Detector Design

Floating strip readout anodes formed by bare copper strips with 35 μm thickness have proven to be discharge tolerant and show up to now no signs of aging.

The commissioning and repair of non-bulk Micromegas, i.e. in which the mesh is not permanently attached to the readout structure, is simplified due to the possible opening and cleaning of the amplification region.

For minimum ionizing particle tracking detectors, the readout structure and the cathode plane are carried by FR4 printed circuit boards. If the detectors are foreseen for operation with elevated gas pressure with respect to the surrounding, a deformation of the outer gas-tight skin cannot be completely avoided. The readout structure and cathode plane can be reinforced by light but stable FR4-aluminum honeycomb panels, that lead to an at least smooth deformation. Another approach is to add an additional gas-tight housing outside of readout structure and cathode plane, as it has e.g. been realized for the cathode in the $6.4 \times 6.4\,cm^2$ floating strip Micromegas doublet.

In order to enable single particle tracking in small Micromegas at flux densities larger than $7\,MHz/cm^2$, the readout structure can be further subdivided into pixels.

The minimum signal length can be reduced by adapting the amplification gap width. Reduction to minimum values on the order of 75 μm seems feasible. The production accuracy of the mesh supporting structure, correlated to the gas gain homogeneity, makes a further reduction difficult. A reduction of the drift gap width also decreases the signal length and reduces efficiency to background radiation due to a smaller active volume at the expense of larger energy loss straggling and a reduced pulse height due to lower ionization of traversing charged particles.

For detection of particle tracks, that are almost parallel to the readout structure, a considerable increase of the drift gap into the Time-Projection-Chamber regime is possible. This would allow for a resolution of events with large hit multiplicity, provided that the readout structure is highly segmented into pads or pixels.

7.8.2 Detector Operational Parameters

The discharge probabilities and the signal length can be reduced by using a different gas mixture, based on light noble gases such as helium or neon (Derré and Giomataris 2001; Fonte et al. 1997). The ion drift time, that defines the minimum signal rise time, is thus intrinsically smaller, see Table 2.2. This has furthermore a positive effect on the discharge probability as neutralization of ions at the mesh competes with ion production, that can result in streamer development due to charge densities exceeding $2 \times 10^6 \, e/0.01 \, mm^2$. By addition of gas admixtures like Carbon Dioxide, Methane, Isobutane or Tetrafluromethane, the electron drift velocities can be strongly influenced.

7.8.3 Readout Electronics

The Gassiplex based readout electronics (Sect. 4.1.1) runs stably in combination with Micromegas and provides reliable charge signal information for all detector strips. The currently achievable readout rate on the order of several kHz could be improved by a hardware upgrade to several 10 kHz. The relatively long shaping time matches the typical signal rise time well.

The APV25 based Scalable Readout System (Sect. 4.1.2) provides information about the temporal evolution of charge signals in steps of 25 ns. The maximum readout rate on the order of 800 Hz can be considerably improved by using on-board common mode noise correction, baseline discrimination and signal analysis on the Front End Concentrator card. Basic algorithms for this exist and are further improved in the near future (Zibell 2014). Correct tuning of APV25 shaping time can improve the pulse height of charge signals at low drift fields.

Novel readout ASICs can enable high-rate readout in medical ion tracking applications. If the electric drift field, that defines the signal rise time at small fields, cannot be adapted, e.g. by high-voltage limitations in Time-Projection-Chambers, read out with Micromegas, the integration and shaping time of the preamplifier electronics must be adapted to the signal rise time. For large area Micromegas detectors, the strip capacitance can reach considerable values, such that the preamplifier shaping time should be accordingly large.

References

Bortfeldt J (2010) Development of micro-pattern gaseous detectors—micromegas. Diploma thesis, Ludwig-Maximilians-Universität München. http://www.etp.physik.uni-muenchen.de/dokumente/thesis/dipl_bortfeldt.pdf

CAEN S.p.A. (2010) Technical information manual mod. V1729. http://www.caen.it,

Derré J, Giomataris I (2001) Spatial resolution and rate capability of MICROMEGAS detector. Nucl Instrum Methods A 461(1–3):74–76, ISSN 0168–9002. (8th Pisa Meeting on Advanced Detectors) doi:10.1016/S0168-9002(00)01171-2. http://www.sciencedirect.com/science/article/pii/S0168900200011712

Derré J, Giomataris Y, Zaccone H, Bay A, Perroud J-P, Ronga F (2001) Spatial resolution in micromegas detectors. Nucl Instrum Methods A 459(3):523–531. ISSN 0168–9002. doi:10.1016/S0168-9002(00)01051-2. http://www.sciencedirect.com/science/article/pii/S0168900200010512

Fonte P, Peskov V, Ramsey BD (2013) Streamers in MSGC's and other gaseous detectors. ICFA Instr Bull 15:1. http://www.slac.stanford.edu/pubs/icfa/

Kuger F (2013) Simulationsstudien und Messungen zu Gasverstärkungsprozessen in Micromegas für den Einsatz im ATLAS NewSmallWheel. Master's thesis, Julius-Maximilians-Universität Würzburg

Lippert B (2012) Studien zur Signalentstehung und Parametrisierung der Gasverstärkung in einem Micromegas-Detektor. Bachelor's thesis, Ludwig-Maximilians-Universität München. unpublished, 2012. day-to-day supervision by Bortfeldt J

Thers D, Abbon Ph, Ball J, Bedfer Y, Bernet C, Carasco C, Delagnes E, Durand D, Faivre J-C, Fonvieille H, Giganon A, Kunne F, Le Goff J-M, Lehar F, Magnon A, Neyret D, Pasquetto E, Pereira H, Platchkov S, Poisson E, Rebourgeard Ph (2001) Micromegas as a large microstrip detector for the COMPASS experiment. Nucl Instrum Methods A, 469(2):133–146. ISSN 0168–9002. doi:10.1016/S0168-9002(01)00769-0. http://www.sciencedirect.com/science/article/pii/S0168900201007690

Zibell A (2014) High-Rate Irradiation of 15 mm muon drift tubes and development of an ATLAS compatible readout driver for micromegas detectors—in preparation. PhD thesis, Ludwig-Maximilians-Universität München

Chapter 8
Summary

In this thesis the development and the performance of novel floating strip Micromegas tracking detectors are discussed. Floating strip Micromegas are versatile, high-rate capable micro-pattern gaseous detectors, with good spatial resolution and detection efficiency. A several millimeter wide drift region and an approximately 0.1 mm wide amplification region are separated by a micro-mesh. Electrons, produced in ionization processes by particles traversing the low field region, drift into the region between the micro-mesh and the anode readout structure, where a high field leads to signal amplification in gas avalanches. Micromegas are intrinsically high-rate capable due to short ion drift paths and a highly segmented readout structure.

High charge densities, produced by strongly ionizing particles, can lead to formation of streamers and subsequent discharges between the micro-mesh and the readout structure. Discharges are non-destructive and do not permanently alter the detector performance, but lead to efficiency drops due to the necessary restoration of the amplification field.

The impact of discharges is considerably reduced in floating strip Micromegas by supplying the copper anode strips individually with high-voltage via high-ohmic resistances. Signals are decoupled over small capacitances. Due to this concept, discharges in floating strip Micromegas are localized and affect only few strips, thus, the dead time is small. The discharge induced efficiency drop is by three orders of magnitude smaller than in standard Micromegas. The microscopic structure of discharges in floating strip Micromegas is investigated in detail and is quantitatively explained with a detailed detector simulation, considering the capacitances within the detector.

Three different detectors are constructed, commissioned and investigated: A $6.4 \times 6.4 \, cm^2$ detector with exchangeable SMD capacitors and resistors allows for an optimization of the floating strip configuration. A floating strip Micromegas doublet of the same dimensions with low material budget is constructed for low-energy ion tracking in medical imaging applications. A second layer of readout strips is added below the anode strips for signal decoupling, avoiding the need for SMD capacitors. A fully integrated $48 \times 50 \, cm^2$ floating strip Micromegas with printed recharge resistors is build for high-energy pion and muon tracking.

© Springer International Publishing Switzerland 2015

J. Bortfeldt, *The Floating Strip Micromegas Detector*, Springer Theses,

DOI 10.1007/978-3-319-18893-5_8

Several analysis algorithms e.g. for track reconstruction, single plane track inclination reconstruction or determination of the spatial resolution, are developed and implemented.

For detector investigation in high-energy particle beams, a tracking telescope, consisting of four layers of $9 \times 10\,cm^2$ standard Micromegas and two layers of $9 \times 9\,cm^2$ resistive strip Micromegas with an intrinsic resolution of each detector on the order of $50\,\mu m$ is developed. The telescope is optimized and employed in high-energy pion and muon beams. Particle track measurements with an accuracy better than $20\,\mu m$ are achieved.

A large area $48 \times 50\,cm^2$ floating strip detector, with 1920 anode strips and a strip pitch of $250\,\mu m$, is investigated in $120\,GeV$ pion beams for homogeneity of its performance. The pulse height behavior is studied, variations below $20\,\%$ show the good homogeneity. An optimum spatial resolution of $(49 \pm 2)\,\mu m$ homogeneously over the detector is observed. The detection efficiency is up to 0.95. Inclination of the detector with respect to the beam allows for an investigation of the track angle reconstruction in a single detector plane. Optimum precision of $\left(^{+4}_{-3}\right)^\circ$ is reached for a track inclination of 40°. Probabilities for particle induced discharges on the order of 10^{-4} are observed.

Cosmic muon tracking measurements in a $6.4 \times 6.4\,cm^2$ floating strip Micromegas under lateral irradiation with $20\,MeV$ proton beams at 550×10^3 protons/s allows for an investigation of the performance in high-rate background environments. The spatial resolution and detection efficiency for cosmic muons are only affected by the proton background irradiation, if the same strips are simultaneously hit as by the cosmic muon. Discharge probabilities per incident proton are 3×10^{-7}, the efficiency is not affected by discharges.

The backflow of positive ions from gas amplification processes into the drift region, can limit the high-rate capability of Micromegas with large drift gaps. It is measured in a resistive strip Micromegas under irradiation with an intense $20\,MeV$ proton beam. At typical working points, the fraction of ions, flowing back into the drift space, is below $2\,\%$.

The floating strip Micromegas doublet is used in two applications that allow furthermore for a characterization of the detector: A proof of principle study for tracking of $20\,MeV$ protons at $550\,kHz$ demonstrates the high-rate tracking capabilities of the detector. Efficiencies above 0.99 are reached. Measurements with tilted detectors with respect to the beam allow for a detailed study and an optimization of the single plane track angle reconstruction method. A detailed detector simulation is developed, that quantitatively explains the observed systematic deviations by capacitive coupling of adjacent strips and enables a correction of systematic deviations. Optimum resolution of $\left(^{+3}_{-2}\right)^\circ$ are observed for a track inclination of 40°.

The good multi-hit and temporal resolution of the floating strip Micromegas doublet enables characterization studies of therapeutic proton and carbon ion beams at the Heidelberg Ion Therapy center. The microscopic temporal structure of the beam is resolved, highly efficient detection of particles up to the highest available rate of $2\,GHz$ is possible. The pulse height decreases by only $20\,\%$ for a rate increase from 2 to 80 MHz, due to the low internal charge-up. Single particle tracking is possible at particle flux densities up to $7\,\text{MHz/cm}^2$. The spatial resolution for carbon ions at a rate of 5 MHz in a single detector layer is better than 180 μm for all available particle energies, which is completely sufficient for particle tracking in imaging applications. The spatial resolution is limited by the 0.5 mm strip pitch of the detectors. Despite of discharge rates on the order of several 10 Hz at GHz particle rates, the detector uptime is above 0.99. Rate independent discharge probabilities on the order of 5×10^{-7} are observed up to the highest particle rates.

This thesis shows, that Micromegas in general and floating strip Micromegas in particular can serve a great breadth of applications. Low material budget floating strip Micromegas are foreseen for ion tracking in medical ion range imaging. A combined system of several Micromegas and a scintillator based range telescope is currently under development.

Large dimensional Micromegas of square meters size are foreseen for the upgrade of the Small Wheel of the ATLAS muon spectrometer. These will exploit both the temporal resolution with angle reconstruction for triggering and the high spatial resolution for precision tracking. Algorithms and methods developed in the context of this thesis, such as the correction of systematic effects in the single plane angle reconstruction, are adopted in the Muon ATLAS Micromegas Activity collaboration (MAMMA). The knowledge about Micromegas, acquired by this thesis indispensably contributes to the design optimization, and construction of the resistive strip Micromegas to be built for the New Small Wheel of ATLAS.

Appendix A
Readout Electronics and Services

A.1 Gassiplex Based Readout System

A Gassiplex based readout system is used to acquire charge signals in Micromegas (Sect. 4.1.1).

The applied preprocessing front-end boards (Fig. A.1) have originally been developed for readout of the cathode plane of the HADES[1] Ring Imaging Cerenkov detector (Kastenmüller et al. 1999). The analog circuit on the boards was modified to cope with the negative charge signals encountered in Micromegas. Each front-end module carries four 16 channel Gassiplex chips. Analog-to-digital conversion, digital threshold comparison and multi-event buffering is performed on the front-end boards, that are directly mounted on the detectors. The on-board signal processing and control of the Gassiplex chips is managed by a Xilinx XC4000E Field Programmable Gate Array (FPGA).

Module control and data handling is performed through a custom-built VME based readout controller, that can serve up to 48 modules in eight groups of six.[2] The readout controller communicates via the VME private bus with the Detector Trigger Unit, which accepts the NIM trigger signal and performs front-end module and readout controller busy handling. The readout program runs on a CES[3] RIO2 VME controller, which is equipped with an embedded Power PC, providing fast access to the VME modules over the VME bus. Additional VME CAEN V775 Time-to-Digital converters were added (CAEN S.p.A. 2012), acquiring signals from triggering scintillators and a 12 bit event counter, that was used to merge data streams from different readout systems, see Sect. 4.1.3. A detailed description of the necessary modifications and the different components on the front-end modules as well as essential grounding considerations can be found in Bortfeldt (2010, Chap. 5).

---

[1] High Acceptance DiElectron Spectrometer.

[2] With modified front-end back plane 64 modules in total are possible.

[3] Creative Electronic Systems.

© Springer International Publishing Switzerland 2015
J. Bortfeldt, *The Floating Strip Micromegas Detector*, Springer Theses,
DOI 10.1007/978-3-319-18893-5

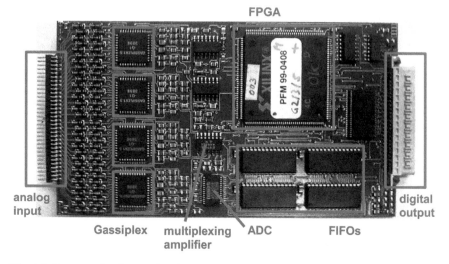

Fig. A.1 Preprocessing front-end module, carrying four charge sensitive Gassiplex chips. Analog signals from the four Gassiplex chips are multiplexed into the 10 bit ADC and stored after digitization in the FIFOs

Several front-end modules are daisy chained and communicate with the readout controller via a backplane, which also provides mechanical stability. In the original HADES design, backplanes supporting four or five modules have been used. Since this electronic was meant to be used with 360 strip Micromegas, new backplanes carrying six front-end modules have been designed. In order to immunize the whole readout chain against low voltage drops after discharges in the detectors, it was essential to buffer the low voltage lines on the backplane with low-ESR[4] capacitors of different capacitances.

A schematic drawing of the readout system and the trigger logic in Micromegas calibration measurements with high-energy pions and muons (Sect. 5.1) can be found e.g. in Fig. 5.1.

A.2 APV25 Based Scalable Readout System

APV25 based front-end boards, interfaced with the Scalable Readout System (Martoiu et al. 2013) are used for acquiring time resolved charge signals in Micromegas detectors (Sect. 4.1.2). The scalable readout system has been developed in the framework of the RD51 collaboration at CERN (Pinto 2010).

---

[4]Equivalent Series Resistance.

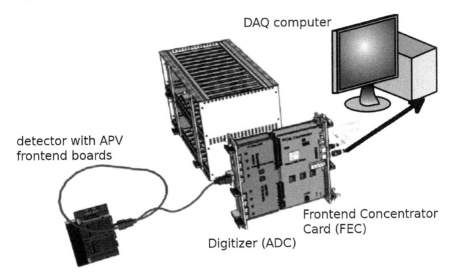

Fig. A.2 Schematic drawing of the SRS with APV25 readout, adapted from (Toledo et al. 2011)

A drawing of the smallest table top version consisting of APV25 front-end boards, a digitizer card, a Frontend Concentrator Card and the Data Acquisition Computer can be seen in Fig. A.2.

Specifically designed hybrid[5] front-end boards are equipped with a single APV25 chip, Fig. A.3. Connection to the detector is realized over 130 pin Panasonic connectors, all channels are AC coupled and protected against discharges via fast diodes. A so called master hybrid additionally contains the PLL25 ASIC (Placidi et al. 2000), which disentangles the 40 MHz bunch clock and the trigger signal, that are fed to the hybrid board over a single differential line pair. An occurring trigger is marked by a missing clock tick. The clock and trigger signals can be passed to a so called slave hybrid such that control and parallel readout of two front-end boards over a single HDMI[6] cable is possible.

The front-end boards are connected to Digitizer or ADC-cards, that contain i.a. two octal 12 bit 40 MHz ADCs for digitization of the raw analog data and have been specifically designed for interfacing up to 16 APV25 hybrids. The ADC-card is mounted via PCI connectors on the generic Frontend Concentrator Card (FEC), which, being equipped with a Xilinx Virtex 5 FPGA, provides readout flow control as well as digital baseline suppression and data preprocessing. The interface to the Data Acquisition (DAQ) Computer is realized via fiber or copper based Gigabit Ethernet. Raw data is transmitted as one over-sized UDP frame per channel,[7] containing the

---

[5]This term is used within the RD51 collaboration and refers to the mixed analog and digital signals on the front-end boards.

[6]High Definition Multimedia Interface.

[7]User Datagram Protocol. Its specifications allows for frames with maximum size of 1518 bytes, communication with larger frames, so called Jumbo frames is often possible.

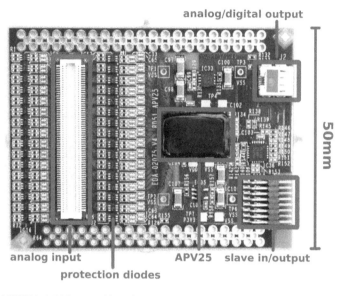

analog/digital output

50mm

analog input   APV25   slave in/output

protection diodes

Fig. A.3 APV25 hybrid front-end board

data of all requested time bins. The typical size for one frame is about 4 kB. For readout of systems with more than 2048 channels or within e.g. the complex ATLAS readout chain, the FEC also exhibits a Data, Timing and Control interface, allowing for the common readout of up to 40 FECs over a so called Scalable Readout Unit (Zibell 2014).

Due to the synchronous transmission of the asynchronous trigger to the APV25 frontend boards, the recorded time resolved charge signals are afflicted with a 25 ns jitter. In applications, where the absolute signal timing is needed, the jitter can be corrected.

Three methods are in principle available: First, directly measure the delay between the analog trigger signal and the synchronous trigger to the APV25 with an external Time-to-Digital converter (Lösel 2013). This yields accurate results, requires a second readout electronics though, that has to be synchronized with the APV25 readout.

Second, record the analog trigger signal or a signal with a fixed delay to the trigger with an APV25 front-end board (Sect. 4.1.3). The trigger timing can then be extracted by fitting the recorded signal with an appropriate function (Sect. 4.3). The accuracy of the correction is on the order of ∼5 ns and depends on the quality of the signal fit. It can be improved by recording several slightly shifted trigger signals on separate channels.

Third, when two inclined, oppositely oriented detectors are available, the time jitter will shift the two separately reconstructed tracklets into opposite directions. By reversing this shift, the time jitter can be determined.

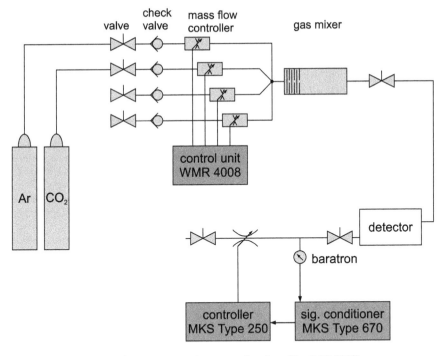

Fig. A.4 Gas mixture and pressure control system, taken from (Bortfeldt 2010)

A.3 Flexible System for Gas Mixing and Pressure Control

A flexible system is used to mix desired detector gases from up to four constituent gases, control the pressure in the detector system and regulate the gas flux.

The applied system has already been described by Bortfeldt (2010) and is again mentioned here for reference. A schematic can be seen in Fig. A.4.

The ratio of constituent gases is measured and controlled by up to four Brooks Thermal Mass Flow Controllers.[8] The constituent gases are mixed turbulently in a passive multi-disk mixture cylinder, developed at LMU, and then passed to the detectors, which are connected in series. By measuring the pressure at the outlet of the detector system[9] and comparing it to a preset value in a PI-controller,[10] an electronically regulated exhaust valve[11] is opened or closed to stabilize the pressure within the detector system.

---

[8] Two 5850S (2008a) and two SLA5850 (2008b) mass flow controllers.

[9] An Absolute High Accuracy Pressure Transducer Type 690 A from (1997b) with an accuracy of $\pm 0.08\%$ reading in combination with a MKS Instruments Signal Conditioner Type 670B is used to measure the pressure.

[10] Type 250 Pressure/Flow Control Module from (1997a).

[11] Flow Control Valve Type 248 A from (1997a).

References

Bortfeldt J (2010) Development of micro-pattern gaseous detectors—micromegas. Diploma thesis, Ludwig-Maximilians-Universität München. http://www.etp.physik.uni-muenchen.de/dokumente/thesis/dipl_bortfeldt.pdf

Brooks Instrument (2008a) Brooks smart (DMFC) MFC/MFM. http://www.brooksinstruments.com

Brooks Instrument (2008b) Brooks models SLA5850, SLA5851, SLA5853 mass flow controllers and models SLA5860, SLA5861, SLA5863 mass flow meters. http://www.brooksinstruments.com

CAEN S.p.A. (2012) Technical information manual mod. V775 & V775N. http://www.caen.it

Kastenmüller A, Böhmer M, Friese J, Gernhäuser R, Homolka J, Kienle P, Körner H-J, Maier D, Münch M, Theurer C, Zeitelhack K (1999) Fast detector readout for the HADES-RICH. Nucl Instrum Meth A 433(1–2):438–443. doi:10.1016/S0168-9002(99)00321-6. http://www.sciencedirect.com/science/article/B6TJM-3X64HB7-26/2/ab4be60e608f86382fea108ceb2ed8be. ISSN 0168-9002

Lösel P (2013) Performance studies of large size micromegas detectors. Master's thesis, Ludwig-Maximilians-Universität München.http://www.etp.physik.uni-muenchen.de/dokumente/thesis/master_ploesel.pdf

Martoiu S, Muller H, Tarazona A, Toledo J (2013) Development of the scalable readout system for micro-pattern gas detectors and other applications. JINST 8(03):C03015. http://stacks.iop.org/1748-0221/8/i=03/a=C03015

MKS Instruments (1997a) MKS Type 248A. http://www.mksinst.com

MKS Instruments (1997b) MKS Type 690A Absolute High Accuracy Pressure Transducer. http://www.mksinst.com

Pinto SD (2010) Micropattern gas detector technologies and applications, the work of the RD51 collaboration. In: Nuclear science symposium conference record (NSS/MIC), 2010 IEEE, pp 802–807. doi:10.1109/NSSMIC.2010.5873870

Placidi P, Marchioro A, Moreira P (2000) CMS tracker PLL reference manual. CERN, Geneva Switzerland. http://cds.cern.ch/record/1069705/files/cer-002725460.pdf

Toledo J, Muller H, Esteve R, Monzó JM, Tarazona A, Martoiu S (2011) The Front-end concentrator card for the RD51 scalable readout system. JINST 6(11):C11028

Zibell A (2014) High-rate irradiation of 15 mm muon drift tubes and development of an atlas compatible readout driver for micromegas detectors—in preparation. PhD thesis, Ludwig-Maximilians-Universität München

Appendix B
Mathematic Methods and Algorithms

B.1 Hough Transform Algorithm

The Hough transform algorithm, proposed by Hough (1959) for the automated analysis of bubble chamber pictures, can be used to find lines or patterns in pixel images (Duda and Hart 1972). This is equivalent to finding tracks in a set of measured hit positions.

In this thesis, the Hough transform algorithm is used for track reconstruction in high hit multiplicity environments (Sect. 4.42) and for noise suppression in the so called μTPC-reconstruction (Sect. 4.8).

The principal idea is illustrated in Fig. B.1. A point (z_i, x_i) in two-dimensional position space can be interpreted as parameters for a function $a(b)$ in the so called Hough space. The transformation function can e.g. be a straight line

$$a(b) = z_i b + x_i , \tag{B.1}$$

where z_i is the slope and x_i the intersect. By inverting Eq. (B.1), we see, that an arbitrary point on the line in Hough space $(b_j, a_j) \in a(b)$ defines a line in position space

$$x(z) = -b_j z + a_j , \tag{B.2}$$

that crosses the initial point (z_i, x_i) by definition.

Generally speaking, the points on the line in Hough space define the family of all lines in position space that intersect with the original point.

Considering now our example in Fig. B.1, we transform six points in position space into six lines in Hough space. The point of intersection of four lines (b_1, a_1) defines the line in position space that intersects with all four points and is thus the searched track.

In order to avoid problems with steep tracks, where a or b can become very large, in practice we use the Hesse normal form line representation as transformation function

© Springer International Publishing Switzerland 2015
J. Bortfeldt, *The Floating Strip Micromegas Detector*, Springer Theses,
DOI 10.1007/978-3-319-18893-5

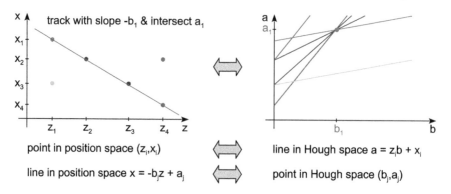

Fig. B.1 Transform of six points in position space (*left*) into six lines in Hough space (*right*). The intersection point of the *dark green, blue, dark magenta* and *brown line* in Hough space defines the parameters of the *red* track, crossing the respective points in position space

$$r(\alpha) = x_i \cos(\alpha) + z_i \sin(\alpha) , \tag{B.3}$$

where $r(\alpha)$ is the shortest distance of the line to the origin and α the angle between line and z-axis in position space.

Since usually the points in position space do not lie not exactly a straight line due to either an inaccurate measurement or due to physical scattering, the transformed functions in Hough space do not intersect in a single point. In order to automatically extract the track parameters anyway, two different methods are used.

First, we can construct a circle with a radius of the same order as the point accuracy around each in position space, and transform all points within this circle into Hough space. The Hough space is modeled by a two-dimensional histogram, whose bins are then filled according to the set of transform functions. The most probable line parameter set is then given by the bin with the maximum number of entries, see Fig. 4.23. This method is universal and is used in the μTPC-reconstruction for eliminating noise or mis-measured timing values, correct tuning of the circle radius though is necessary, see Sect. 4.8.2.

Second, we can analytically calculate the points of intersection of the Hough transform functions in Hough space and cluster them in a physical correct way, until the correct number of intersection points is combined, see Sect. 4.4.2. This method produces very reliable results, but can in the present implementation only be used in track reconstruction, since it is explicitly assumed, that each detector layer contributes with one cluster to the track.

B.2 Analytic Line Fit and its Uncertainty

As it has been discussed by Horvat (2005), from where the content of the following section is taken, the fit of a two parameter function such as a straight line $x(z) = mz + a$ to a set of data points (z_i, x_i) can be performed analytically. This can be used for e.g. track determination from a set of measured hit position points or for reconstruction of a tracklet from measured time-strip points in the single plane angular reconstruction.

Fitting the function $x(z)$, determined by the two parameters m and a, to a set of i points with the uncertainty w_i for each point, is equivalent to minimizing the sum

$$\chi^2 = \sum_i \frac{(mz_i + a - x_i)^2}{w_i^2} \tag{B.4}$$

with respect to a and m:

$$\frac{\partial \chi^2}{\partial a} = \sum_i \frac{2(mz_i + a - x_i)}{w_i^2} = 0$$
$$\frac{\partial \chi^2}{\partial m} = \sum_i \frac{2z_i(mz_i + a - x_i)}{w_i^2} = 0 . \tag{B.5}$$

We can now define coefficients g_1, g_2, Λ_{11}, Λ_{12} and Λ_{22} as

$$g_1 := \sum_i \frac{x_i}{w_i^2}, \quad g_2 := \sum_i \frac{x_i z_i}{w_i^2}$$
$$\Lambda_{11} := \sum_i \frac{1}{w_i^2}, \quad \Lambda_{12} := \sum_i \frac{z_i}{w_i^2} \text{ and } \Lambda_{22} := \sum_i \frac{z_i^2}{w_i^2}. \tag{B.6}$$

Inserting these coefficients into Eq. (B.5) yields

$$\Lambda_{11}a + \Lambda_{12}m = g_1$$
$$\Lambda_{12}a + \Lambda_{22}m = g_2 . \tag{B.7}$$

This can be solved for intersect a and slope m using Cramer's rule

$$a = \frac{g_1\Lambda_{22} - \Lambda_{12}g_2}{D} \text{ and}$$
$$m = \frac{\Lambda_{11}g_2 - g_1\Lambda_{12}}{D} , \tag{B.8}$$

where

$$D := \Lambda_{11}\Lambda_{22} - \Lambda_{12}\Lambda_{12}. \tag{B.9}$$

As discussed in Sect. 4.5.3, a reliable method to measure the unknown spatial resolution of a detector is to interpolate a track, predicted by detectors with known resolution, into the detector under test. By comparing the predicted with the measured particle hit position, the spatial resolution can be calculated, using Eq. (4.16). Therefor the track accuracy σ_{track} has to be known.

Assume a large number of identical tracks, measured by n detectors with known spatial resolutions σ_i. Due to the finite accuracy of the reference detectors, each track is reconstructed slightly differently. The predicted hit position $x_{\text{pred}}(z)$ scatters around its expectation value $\langle x_{\text{pred}}(z) \rangle$. The track accuracy $\sigma_{\text{track}}(x, z)$ is thus given by

$$\sigma_{\text{track}}^2(x, z) = \langle (x_{\text{pred}}(z) - \langle x_{\text{pred}}(z) \rangle)^2 \rangle$$
$$= \text{Var}(a) + 2z\text{Covar}(a, m) + z^2\text{Var}(m) . \tag{B.10}$$

A detailed derivation of the variance and covariance can be found in (Horvat 2005). Equation (B.10) ultimately leads to a relation between the track accuracy and the spatial resolutions σ_i of the involved detectors:

$$\sigma_{\text{track}}(x, z) = \sqrt{\frac{\Lambda_{22} - 2z\Lambda_{12} + z^2\Lambda_{11}}{D}} . \tag{B.11}$$

The single detector spatial resolutions $\sigma_i = w_i$ are interpreted as the weights w_i, used in Eq. (B.6).

B.3 Histogram Binning for Transformed Variables

Histograms often represent a viable mechanism to display or analyze variables, being distributed according to complex distribution functions.

Consider a quantity x, distributed according to the probability density function $f_x(s)$. Often other quantities $g(x)$ are deduced from x. This derived quantity is then distributed according to

$$f_g(s) = \left| \frac{dg^{-1}(s)}{ds} \right| f_x(g^{-1}(s)) . \tag{B.12}$$

To give an example x may be the reconstructed track slope, derived with the μTPC method, Sect. 4.8, and the deduced variable is in this case the track inclination angle $\vartheta = g(x) = \arctan(1/x)$. If a histogram is used to represent the distribution of the reconstructed track angles, a systematic error is introduced when the histogram has a fixed bin size.

This can be understood mathematically, since the physical relevant quantity is the probability of finding a track with a certain slope or with a certain track inclination angle respectively, which must be unaffected by the transformation:

$$P(a \leq x \leq b) = \int_a^b f_x(s)ds = \int_{g(b)}^{g(a)} f_g(s)ds = P(g(b) \leq g(x) \leq g(a)) . \tag{B.13}$$

Fig. B.2 Random variable
x, distributed according to
Gaussian distribution
functions with common
mean 0.7 and standard
deviations 0.1 (*black*), 0.2
(*red*) and 0.3 (*green*)

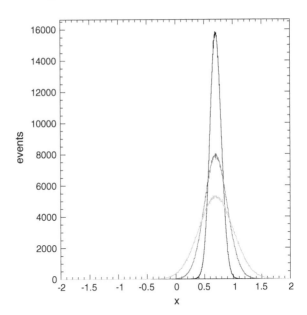

Note the commutation of the integral bounds, which is necessary for a transform
function with $dg(x)/dx < 0$.

An example is shown in Figs. B.2 and B.3. The randomly distributed variable x
follows a Gaussian probability density function with mean $\bar{x}$. The most probable value
of $\arctan(1/x)$ is obviously expected to be $\arctan(1/\bar{x})$ as the arctan is a monotonous
function. As it can be seen in Fig. B.3 though, the most probable value is shifted
towards lower values for increasing width of the original Gaussian distribution. This
unphysical behavior is caused by the incorrect binning of the histogram with a fixed
bin width.

Going back to our application, this would lead to an unphysical dependence of
the reconstructed track angle ϑ on the shape of the distribution of the track slopes
x. In order to avoid this subtle, but devastating effect, a correct transformation of
$x \rightarrow g(x)$ involves the mutual transformation of the bin width $dt = t_{i+1} - t_i \rightarrow$
$|g(t_{i+1}) - g(t_i)| = ds$. For any non-linear transformation, this leads to the necessity
of variable bin width for the transformed histogram.

In Fig. B.4, the distribution for $\arctan(1/x)$ is displayed in a histogram with correctly transformed variable bin width. The most probable value is now indeed independent of the x-distribution.

It should be noted, that the bin size of the transformed histogram tends to 0 for
$\arctan(1/x) \rightarrow 0$, i.e. the number of bins diverges. For the displayed case, the bin
width between $\arctan(-1/5)$ and $\arctan(1/5)$ has been set to a constant value. As
long as no physically relevant values lie in this region, this is acceptable as it enables
the representation of the whole possible value range in a single histogram.

Fig. B.3 Transformed
variable arctan($1/x$), where
the distribution for x is
shown in Fig. B.2. The
shown histogram has a fixed
bin width. Clearly visible is
the unphysical dependence
of the most probable
transformed value on the
width of the distribution of x

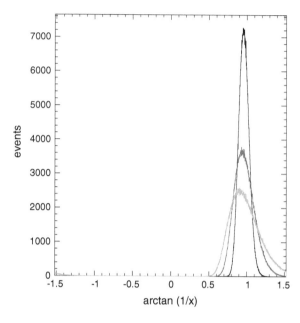

Fig. B.4 Transformed
variable arctan($1/x$), where
the distribution for x is
shown in Fig. B.2. The
shown histogram has a
variable bin width

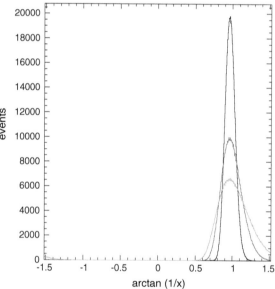

B.4 Probability for a Random Background Hit Within an Acquisition Window

Consider an experiment with a detector system, in which the data acquisition is triggered by traversing particles. Upon a trigger, the data acquisition is recording detector signals for a fixed time window with length T. Also consider a random background irradiation with mean hit rate f. The probability of finding k background hits within the time window is given by the Poisson distribution

$$P(k) = \frac{\lambda^k}{k!} e^{-\lambda} , \tag{B.14}$$

where $\lambda = Tf$ is the mean number of expected background hits in the time window.

Thus the probability of finding at least one background hit in the triggered event is given by

$$P(k \geq 1) = 1 - P(0) . \tag{B.15}$$

References

Duda RO, Hart PE (1972) Use of the hough transformation to detect lines and curves in pictures. Commun ACM 15(1):11–15

Horvat S (2005) Study of the higgs discovery potential in the process $pp \rightarrow H \rightarrow 4\mu$. PhD thesis, Zagreb University

Hough PVC (1959) Machine analysis of bubble chamber pictures. In: Proceedings of the international conference on high energy accelerators and instrumentation

Appendix C
Supplementary Material—Applications of Floating Strip Micromegas

In the following, supplementary material and detailed discussions of results from applications of a floating strip Micromegas doublet in 20 MeV proton tracking measurements and proton and carbon ion beam characterization measurements can be found.

C.1 Tracking of 20 MeV Protons with Floating Strip Micromegas

A floating strip Micromegas doublet with low material budget has been used for tracking measurements of 20 MeV protons at the tandem accelerator in Garching. The measurement campaign is described in Sect. 6.1.

C.1.1 Angular Resolution

The reconstruction of track inclination in a single detector plane has been investigated, by rotating the detector with respect to the beam. A summary of the inclination reconstruction capabilities can be found in Sect. 6.1.5.

The quality of the angle reconstruction and the shape of the resulting distributions of track inclination can be estimated by determining the fraction of events within certain intervals. In Table. C.1 the fractions are summarized as a function of the angle for the eight points, shown in Fig. 6.12.

© Springer International Publishing Switzerland 2015
J. Bortfeldt, *The Floating Strip Micromegas Detector*, Springer Theses,
DOI 10.1007/978-3-319-18893-5

Table C.1 Fraction of reconstructed track inclinations within certain inclination interval, measured at $E_{amp0} = 33.3\,kV/cm$ and $E_{amp1} = 31.7\,kV/cm$ and a drift field of $E_{drift0,1} = 0.17\,kV/cm$

Angle	$> 2\sigma_<$	$> 1\sigma_<$	$< 1\sigma_>$	$< 2\sigma_>$	in $1\sigma_{<,>}$	in $2\sigma_{<,>}$	$< 2\sigma_<$	$> 2\sigma_>$
Layer 1								
10°	0.321	0.172	0.207	0.296	0.379	0.617	0.227	0.157
20°	0.360	0.236	0.293	0.391	0.528	0.751	0.089	0.163
30°	0.360	0.253	0.294	0.394	0.547	0.754	0.082	0.171
40°	0.399	0.292	0.309	0.420	0.601	0.819	0.071	0.126
Layer 2								
10°	0.295	0.168	0.190	0.279	0.358	0.574	0.252	0.175
20°	0.380	0.247	0.279	0.374	0.526	0.754	0.097	0.153
30°	0.385	0.262	0.282	0.377	0.544	0.762	0.099	0.145
40°	0.417	0.295	0.303	0.404	0.598	0.821	0.078	0.114

The shown values are afflicted with a maximum uncertainty of ± 0.01. Intervals around the most probable reconstructed angle ϑ_{mp} are considered. The angular accuracy for lower angles is $\sigma_<$ and $\sigma_>$ for larger angles, respectively. Also compare to the definition in Sect. 4.8.3. For the calculation of the fractions, the angular accuracies from Fig. 6.12 have been used. $> 2\sigma_<: [\vartheta_{mp} - 2\sigma_<, \vartheta_{mp}]$, $> 1\sigma_<: [\vartheta_{mp} - 1\sigma_<, \vartheta_{mp}]$, $< 1\sigma_>: [\vartheta_{mp}, \vartheta_{mp} + 1\sigma_>]$, $< 2\sigma_>: [\vartheta_{mp}, \vartheta_{mp} + 2\sigma_>]$, in $1\sigma_{<,>}$: inside the one-sigma band around ϑ_{mp}, in $2\sigma_{<,>}$: inside the two-sigma band around ϑ_{mp}, $< 2\sigma_<$: smaller than $\vartheta_{mp} - 2\sigma_<$, $> 2\sigma_>$: larger than $\vartheta_{mp} + 2\sigma_>$

C.1.2 Spatial Resolution

The spatial resolution can be approximated by directly comparing measured hit positions in the two detector layers. A considerable degradation of the observed resolution is caused by the beam divergence, which is considerably enhanced by multiple scattering of the 20 MeV protons in the two readout structures. From runs with nearly equal operational parameters for both detector layers, the common single layer spatial resolution is determined, using Eq. 6.3. The operational parameters of one layer can then be varied, while keeping the other layer's constant, to determine the influence of e.g. the drift and amplification fields on the spatial resolution.

The single layer spatial resolution as a function of the drift field is shown in Fig. C.1 for different amplification fields. As the measured hit residual distribution are considerably broadened by multiple scattering of the incident protons, the spatial resolution appears to be independent of the drift field for the higher gas gains. A weak degradation with increasing drift field is visible for the lower amplification fields as the intrinsic spatial resolution is of the same order as the contribution from multiple scattering. Note that the values are similar in both layers by definition.

In Fig. C.2 the spatial resolution as a function of the drift field for different particle rates and differently shaped beams is shown.

The hit position per layer has up to now been reconstructed with the usual charge-weighted-mean method, described by Eq. 4.2. This yields reliable and robust results for perpendicular and nearly perpendicular tracks. As can be seen in Fig. C.3, the

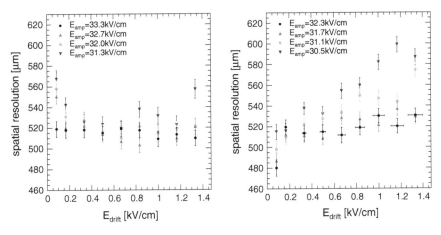

Fig. C.1 Spatial resolution as a function of the drift field for the first (*left*) and the second detector layer (*right*) for different amplification fields, Measured with perpendicularly incident 20 MeV protons at a mean rate of (420 ± 10) Hz

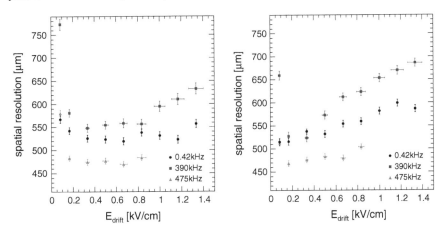

Fig. C.2 Spatial resolution as a function of the drift field for the first (*left*) and the second detector layer (*right*). Measured with different particle rates and beam foci shapes. During the two measurements with lower rates, a prolate beam spot of $(22.4 \pm 0.5 \times 1.3 \pm 0.2)$ mm was used. For the high-rate measurement and the angular scans, the beam was refocused to $(4.5 \pm 0.2 \times 3.5 \pm 0.2)$ mm^2. Measured with amplification fields $E_{amp0} = 31.3$ kV/cm and $E_{amp1} = 30.5$ kV/cm

spatial resolution significantly degrades for both detector layers from 500 μm at 0° and 10° inclination to 1100 μm at 40°.

The dependence of the spatial resolution on the drift field becomes more pronounced with increasing track inclination, as the contribution of the single detector spatial resolution to the width of the hit residual distribution increases significantly, contrary to the multiple scattering contribution.

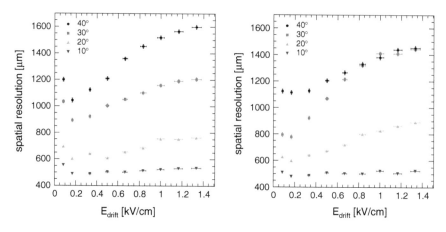

Fig. C.3 Spatial resolution as a function of the drift field for four different track inclinations for the first and the second detector layer (*left* and *right*). Measured at $E_{amp0} = 33.3$ kV/cm and $E_{amp1} = 31.7$ kV/cm. The hit position in each detector layer has been reconstructed with the usual charge weighted mean-method, see Eq. 4.2 and Sect. 4.3

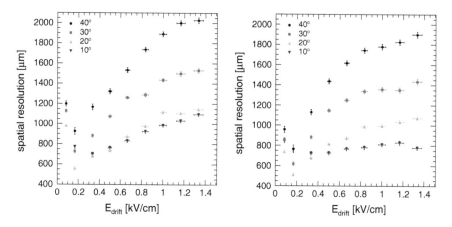

Fig. C.4 Spatial resolution as a function of the drift field for the first (*left*) and the second detector layer (*right*). The hit position in each detector layer is defined by the zero of the time-strip-points fit, used to determine the track inclination in the µTPC-reconstruction method. Measured at $E_{amp0} = 33.3$ kV/cm and $E_{amp1} = 31.7$ kV/cm

For inclined tracks, an alternative hit position is given by the zero of the straight line fit to time-strip-data points, used in the µTPC-reconstruction, see Sect. 4.8.1. The resulting spatial resolution is shown in Fig. C.4.

For track inclinations of 20°, 30° and 40°, an improvement with respect to the charge-weighted-mean reconstruction of 15–20 % is visible at the optimum drift field $E_{drift} = 0.17$ kV/cm. It has been shown in Sect. 6.1.5, that also the angular resolution is best in this field region.

For increasing drift field, the angular resolution and with it the accuracy of the hit position measurement degrades considerably. For the highest drift fields and inclinations of 20°, 30° and 40°, the spatial resolution, when using the μTPC-fit for position reconstruction, is by 30–20 % worse than for the usual method.

C.1.3 Discharge Behavior

Discharges have been counted by registering for each layer separately the recharge signal at the common strip potential. Particle rates were measured by counting signals from the triggering scintillators. The discharge probability is then given by

$$P_d = \frac{n_{\text{discharge}}}{n_{\text{trigger}}}. \tag{C.1}$$

Its error, dominated by the statistic uncertainty of $n_{\text{discharge}}$ can be calculated, assuming Poisson statistics:

$$\Delta P_d = \frac{\sqrt{n_{\text{discharge}}}}{n_{\text{trigger}}}. \tag{C.2}$$

During the measurements with perpendicularly incident, low rate beam, no discharges were observed. The discharge probability for the parameter set, maximizing the number of discharges i.e. 10° track inclination, high amplification fields, high particle rate, is shown in Fig. C.5 as a function of the drift field.

Similar values for both detector layers are observed, with a maximum of $P_d = (5 \pm 2) \times 10^{-7}$ discharges per incident proton at $E_{\text{drift}} = 0.17\,\text{kV/cm}$. This field corresponds also to the maximum electron transmission (Fig. 6.7) and is close to the correlated minimum transverse diffusion (Fig. 2.5). For drift fields $E_{\text{drift}} > 0.5\,\text{kV/cm}$, the discharge probability is compatible with 0.

In Fig. C.6, the dependance on the track inclination angle is shown, for a drift field $E_{\text{drift}} = 0.33\,\text{kV/cm}$ and amplification fields of $E_{\text{amp0}} = 33.3\,\text{kV/cm}$ and $E_{\text{amp1}} = 31.7\,\text{kV/cm}$, since about 18 min measurement time is available for each angle.

The discharge rate in the first layer is larger than in the second, since the proton beam broadens with increasing depth, decreasing the instantaneous flux, due to multiple scattering in the beam window and the detector readout planes. It depends on the track inclination, a decrease by 40 % when going from 10° to 40° is observed. The non-monotonous behavior is probably caused by the variable particle rates between the measurements, leading to an over-proportional variation of the particle flux. The particle rates were (460 ± 10) and (445 ± 10) kHz for the 10°, (420 ± 10) and (360 ± 30) kHz for the 20°, (510 ± 15) and (520 ± 15) kHz for the 30° and (490 ± 10) and (470 ± 10) kHz for the 40° measurements, where the rates during the measurement of P_{d0} were quoted first.

Fig. C.5 Discharge probability as a function of the drift field. Measured with a (470 ± 20) kHz proton beam, inclined by $10°$ at amplification fields $E_{amp0} = 33.3$ kV/cm and $E_{amp1} = 32.7$ kV/cm. Due to the short measurement times, considerable errors must be accepted

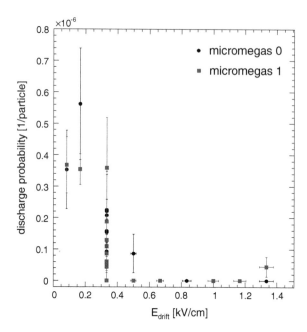

Fig. C.6 Discharge probability as a function of track inclination for the first and the second Micromegas layer. Measured with 20 MeV protons at amplification fields $E_{amp0} = 33.3$ kV/cm and $E_{amp1} = 31.7$ kV/cm and a drift field $E_{drift} = 0.33$ kV/cm. Errors have been calculated, assuming Poisson statistics for the number of discharges per time interval

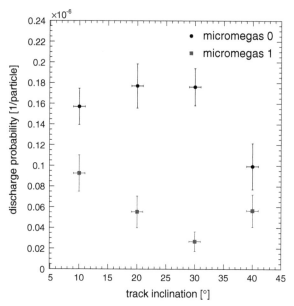

Fig. C.7 Discharge probability per incident particle as a function of the beam energy, measured with carbon ions at $I_{12C} = 5 \times 10^6$ Hz and $E_{amp0} = 30.7$ kV/cm, $E_{amp1} = 29.9$ kV/cm and $E_{drift0,1} = 0.33$ kV/cm

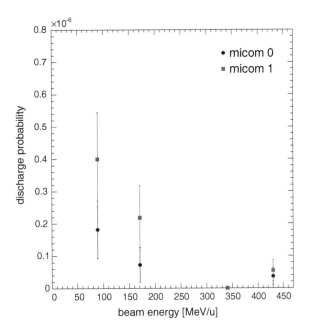

C.2 Characterization of High-Rate Proton and Carbon Beams with Floating Strip Micromegas—Discharge Behavior

A floating strip Micromegas doublet with low material budget has been tested in high-rate proton and carbon ion beams (Sect. 6.2). The discharge probability in carbon beams as a function of the particle energy is shown in Fig. C.7. It decreases with increasing beam energy due to a decreasing ion energy loss in the active region of the detector.

In Fig. C.8, the same quantity is shown for proton beams. The outlier at 91.48 MeV/u is caused by a long spill-pause of about 9 s during the measurement, that leads to an underestimation of the total number of incident particles. The large variance in the spill-pause duration is also visible in Fig. 6.17.

Due to increased amplification fields, the discharge probabilities for proton and carbon beams are now similar. The energy dependance of the discharge probabilities in proton beams seems to be smaller than in carbon beams, but due to the large systematic and statistic and the smaller energy range for the proton beams, no clear statement can be made. The major systematic uncertainties are the low time resolution and the inability to measure the instantaneous particle rate.

In Figs. C.9 and C.10 the discharge probabilities in 88.83 MeV/u carbon and 221.06 MeV/u proton beams is shown as a function of the amplification field. They increase as expected with the amplification fields. The behavior of the two layers is different in carbon beams due to the fragmentation of carbon nuclei.

Fig. C.8 Discharge probability per incident particle as a function of the beam energy, measured with protons at $I_p = 8 \times 10^7$ Hz and $E_{amp0} = 34.7$ kV/cm, $E_{amp1} = 34.1$ kV/cm and $E_{drift0,1} = 0.33$ kV/cm

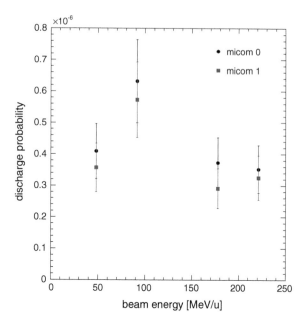

Fig. C.9 Discharge probability as a function of the amplification field, measured with 88.83 MeV/u carbon beams at $I_{^{12}C} = 8 \times 10^7$ Hz and $E_{drift0,1} = 0.33$ kV/cm

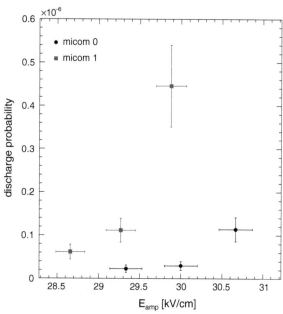

Fig. C.10 Discharge
probability as a function of
the amplification field,
measured with
221.06 MeV/u protons at
$I_p = 2 \times 10^8$ Hz and
$E_{drift0,1} = 0.33$ kV/cm

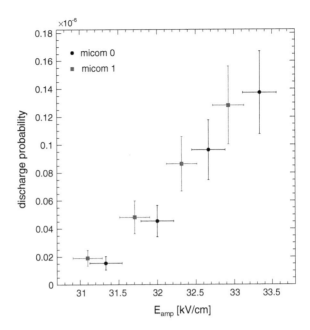

Curriculum Vitae

Jonathan Bortfeldt

Augustenstr. 106
80798 Munich
Germany

Tel.: +49 (0)89 12596738
jonathan@bortfeldt.org

Born July 27, 1985 in Cologne
Nationality: German

Professional Experience

02/2014 to present	**Ludwig-Maximilians-Universität, Munich, Germany**, postgraduate research fellow at the chair for Experimental Particle Physics

Education

10/2004 to 04/2014	**Ludwig-Maximilians-Universität, Munich, Germany**
– 04/2014	Dr. rer. nat. in physics summa cum laude
– 02/2011 to 04/2014	PhD thesis: *Development of Floating-Strip Micromegas Detectors*
– 11/2010	Diploma in physics with distinction (Diplom-Physiker Univ.)
– 11/2009 to 11/2010	Diploma thesis: *Development of Micro-Pattern Gaseous Detectors—Micromegas*
08/2007 to 02/2008	**École Polytechnique Fédérale de Lausanne, Switzerland**, Student exchange with Erasmus Scholarship
06/2004	Abitur at the **Albrecht-Altdorfer-Gymnasium, Regensburg, Germany**
08/2001 to 06/2002	**Horizon High School, Denver, CO, USA**, Student exchange during 11th grade

© Springer International Publishing Switzerland 2015

J. Bortfeldt, *The Floating Strip Micromegas Detector*, Springer Theses,
DOI 10.1007/978-3-319-18893-5

Awards and Funding

11/2014	Munich Excellence Cluster Universe PhD Award 2014 *Experiment*
2014	Seed Money Project of the Munich Excellence Cluster Universe *Three-Dimensional Particle Tracking with Micromegas Detectors at Highest Rates*
02/2011 to 04/2014	PhD thesis funded by a scholarship of the DFG research training group *Particle Physics at the Energy Frontier of New Phenomena*
04/2005 to 10/2010	Scholarship of the Evangelisches Studienwerk Villigst e.V

Teaching

Since 2012	Supervision of three bachelor and two diploma theses
2014 to 2015	Organizing and teaching biweekly tutorials for the Bachelor level particle physics lecture
2014	Teaching Weekly tutorials for the Master level particle physics lecture
2013 to 2014	Teaching biweekly tutorials for the Bachelor level particle physics lecture
2007 to 2009	Teaching physics lab courses for students (physics, advanced physics, biology)

Conference Contributions

07/2014	**International Conference on High Energy Physics, Valencia, Spain**, Presentation: *High-Rate Capable Floating Strip Micromegas Detectors*
03/2014	**Annual conference of the German Physical Society (DPG), Mainz, Germany**, Presentations: *High-Rate Capable and Discharge Tolerant Floating Strip Micromegas Detector* and *High-Rate Capable Micromegas Detectors for Ion Transmission Imaging*
07/2013	**European Phys. Society Conference on High-Energy Physics, Stockholm, Sweden**, Posterpresentation: *Large-Area Floating Strip Micromegas Detectors*
03/2013	**Annual conference of the German Physical Society (DPG), Dresden, Germany**, Presentation: *Performance of a Micromegas Detector with Novel Floating Strip Anode*
05/2012	**Frontier Detectors for Frontier Physics Conference, Portoferraio, Italy**, Posterpresentation: *High-Resolution Micromegas Telescope for Pion- and Muon-Tracking*
03/2012	**Annual conference of the German Physical Society (DPG), Göttingen, Germany**, Presentation: *Behavior of the Spatial Resolution of Micromegas Detectors*
06/2011	**ANNIMA 2011 Conference, Gent, Belgium**, Presentation: *Development of a High-Resolution Muon Tracking Systems Based on Micropattern Detectors*
03/2011	**Annual conference of the German Physical Society (DPG), Karlsruhe, Germany**, Presentation: *Development of a Micromegas based Tracking System*
03/2010	**Annual conference of the German Physical Society (DPG), Bonn, Germany**, Presentation: *Development of High-Resolution Micromesh based Gas Detectors*

Publications

2015	J. Bortfeldt, M. Bender, O. Biebel, H. Danger, B. Flierl, R. Hertenberger, Ph. Lösel, S. Moll, K. Parodi, I. Rinaldi, A. Ruschke, A. Zibell; *High-rate capable floating strip micromegas.* submitted to Nucl. Phys. B, Proc. Suppl
2013	J. Bortfeldt, O. Biebel, R. Hertenberger, Ph. Lösel, S. Moll, A. Zibell; *Large-area floating strip micromegas.* PoS, EPS-HEP2013:061
2013	J. Bortfeldt, O. Biebel, R. Hertenberger, A. Ruschke, N. Tyler, and A. Zibell; *High-resolution micromegas telescope for pion- and muon-tracking.* Nucl. Instr. and Meth. A, 718(0):406–408
2012	J. Bortfeldt, O. Biebel, D. Heereman, R. Hertenberger; *Development of a high-resolution muon tracking system based on micropattern detectors.* IEEE Trans. Nucl. Sci., 59(4):1252–1258
2012 to present	192 publications within the ATLAS collaboration in various journals

CPSIA information can be obtained
at www.ICGtesting.com
Printed in the USA
LVHW081140150320
650074LV00002B/150